RIPPLES IN THE ETHER II

To Colin & Sue
with best wishes.

David C S[illegible]

RIPPLES IN THE ETHER II

DAVID C. SOMERVILLE

This edition published in 2022

Published under licence by Brown Dog Books and
The Self-Publishing Partnership Ltd, 10b Greenway Farm, Bath Rd,
Wick, nr. Bath BS30 5RL

www.selfpublishingpartnership.co.uk

ISBN printed book: 978-1-83952-578-0
ISBN e-book: 978-1-83952-579-7

Cover design by Kevin Rylands
Internal design by Andrew Easton

Printed and bound in the UK

This book is printed on FSC® certified paper

Dedicated to the memory of our dear friend
Raymond Peter Gauld
1934–2018

Many thanks to my wife Marjorie for all her help and support in writing this book.

Thank you to Robin Barrett Astro-photographer 2022 for the use of his astronomical images on the front cover and in the book.

Contents

Preface

The main inspiration for writing *Ripples in the Ether II* comes from a lifelong involvement with the application of frequencies. This started in earnest when I was training as a boy entrant in the Royal Air Force. The trade I trained for was not my first choice but was one offered to me as being more suitable based on the selection process. This was to be in the radio trades group, as an air wireless mechanic. The mysteries of radio transmissions began to unfold, and the methods by which voice information was sent and received were revealed. The basis of all this was centred on the concept of resonant frequencies, and the method of transmission and reception of those frequencies using resonant circuits was, for me, the most intriguing. But this later led me to reason and to realise that all things in life and in existence are full of resonances, not just radio circuits. Such resonances include frequency components in light, heat, and sound, and the things detectable by all the other senses, which provide the resonances in tuned physical and biological circuits as stimuli to our very own existence and the universe as a whole. This extends all the way down to the lowest of the low mechanical resonances, such as those found in nature, from whale communication to earth vibrations.

In my own doctoral research project, the responses of in vivo human long bones to mechanically applied vibrational frequencies played a part, where resonance was one of the characteristics studied to determine the rate of healing of fractures and areas of bone loss. This further proved to me that everything in existence can be broken down and defined by individual resonant qualities. This is just one example of mechanical oscillatory dynamism at the lower end of a vast spectrum of frequencies that defines all aspects of existence. *Ripples in the Ether II* is an attempt to put many aspects of resonances found in our universe across the whole spectrum into context.

Ripples in the Ether II

Stephen Hawking's book *A Brief History of Time* was in a way another inspiration since he starts with the Big Bang and ends with a black hole. I am no Stephen Hawkins and do not have his mathematical or analytical abilities to theorise about space and time, but I do have an enquiring mind and try to reason from my own point of view how things are, where all things began, and how they will eventually end, looking at the bits in between from a dynamic, logical, and in some cases, technical point of view.

If you want to find the secrets of the Universe, think in terms of energy,
frequency and vibration.
Nikola Tesla (1856-1943)

Chapter 1

INTRODUCTORY THOUGHTS

The introduction to many books is generally a short precis of the contents. In *Ripples in the Ether*, the introductory chapter is longer and includes personal recollections and events that are freely used and are set out within this chapter. These helped shape my own thought processes and led to some of the theories about existence and about events and similarities that are found throughout the universe. These theories follow a common theme that relates energy to frequency and frequency to mass. Such oscillations range from those so low to be hardly discernible, except through specific instrumentation, to those so great that, given that the frequencies are of such high oscillatory levels, they are beyond measuring by current scientific sensing devices, although they are detectable by their effects. This introduction is written not to just stimulate interest in the book's contents but also to let the reader see from a personal point of view the grounding for some of the ideas presented. It is not meant to be a scientific textbook; it was written for general interest.

When does science fiction become science fact? Many of the great science fiction writers used their imaginations to create stories about the world and space, envisioning gadgets that seemed far-fetched at the time of their writing but are now in common use. Personal audio and visual communicators are the most obvious, being manifested as 'smart' mobile phones or watches, along with proximity-sensing automatically opening doors. Voice, face, and speech recognition are amongst the many other examples. It could be said that science fiction precedes science fact and that many modern inventions came about after being inspired by science fiction. The human mind has to be ahead of scientific theories and development. Every advancement in science is initially the product of someone's mind.

Theories of how the universe began are also products of the mind and start off as a form of science fiction, having no real reference points on which to base solid facts at the time. However, some fictional ideas suggest a sort of logic that triggers scientific interest, leading to further investigations that, in many cases, become a reality or an accepted theory. Frequency and dynamism are the keys to the realities of some of these and the many other ideas first fictionalised that affect virtually all aspects of modern existence.

This book contains some discussions about reality as we perceive it, but in many cases, the causes and effects are still within the realms of being theoretical. However, fiction, theory, and fact become more intertwined when discussing frequencies since the whole of existence appears to have a frequency component in some form or other. Oscillations create and sustain the universe in its entirety, from those that take billions of years to repeat to other, immeasurably high ones found at the quantum level. These define all aspects of the universe and everything visible and nonvisible, and everything living within the universe. They also provide for a theoretical excursion into how our universe began and how it has progressed and still is progressing.

In writing a book such as this, I am aware that deep particle physics involving mathematical solutions and dimensional analysis of the physics of existence may be beyond the understanding of those who have not specifically studied the subject. Although I have studied some physics and environmental chemistry to a relatively high level, I am not a mathematician and appreciate how complex the maths may appear. However, I have proposed ideas from my own thoughts about the origins of the universe for simplifying the need to understand such complexities without detracting from the deep scientific analysis or the abilities of theoretical physicists, although some simpler equations will be used where necessary within these pages.

As the reader progresses through the book, he or she may note that I have discussed some of the established scientific theories, but there are some presented herein that are largely formulated from my own thoughts and based on experiences gained from my time in military service, industry, and academia. My approach is to try to logically give a simplified, wide-ranging personal point of view to the many questions arising out of this existence and how it came to be, from the Big Bang onwards. I hope this will be seen to have been presented in a readable and understandable way. *Ripples in the Ether II* also has discussions of the nature of frequencies in the electromagnetic spectrum, ranging from gamma radiation to visible light and continuing down to the lowest of frequencies, used in radio communication. The dangers that some of these frequencies may present in terms of direct exposure to both human and animal tissues are also discussed, along with some of the technicalities of design for their uses deriving from both man-made and natural sources. Included are some discussions found in my previously published electrotherapy books, which are cited. *Ripples in the Ether* II is written to be of general interest to all with enquiring minds, especially those who often contemplate existence and our place in the universe. It also includes discussions of electromagnetic frequencies, such as light, X-rays, and heat, along with other frequencies that are used to diagnose illnesses and assist the natural healing processes of the body. This is particularly related to infection, injury, or other pathogenic processes.

Analysis and discussion of frequencies does not stop at the lowest end of the electromagnetic spectrum but continues through the sound spectrum, from ultrasound to the infrasonic. Each stage includes a discussion on how we perceive and make use of some of these frequency bands in medical diagnosis, therapy, and industry. A particular interest of mine focusses upon some of those generated within the human body, including the brain. This includes theories surrounding the mysteries of human consciousness and the ability to be self-aware. The latter part of the book explores the frequencies found within the movement of the planets, the solar system, and extra-terrestrial bodies that can take many millions of years to repeat, thereby being included as amongst the lowest repetitions and, therefore, the lowest frequencies in the universe.

I acknowledge that there may be some who disagree with some of my theories presented herein regarding certain aspects of the universe, from both scientific and religious points of view. Also, perhaps the oversimplification of some of the involved calculations may inspire some critics to comment. However, *Ripples in the Ether* II has been written to provoke discussion about energy and its origins, with discussions on energy in both the early and later chapters. I hope that this work does inform and promote some to think on the subject in a way that they perhaps may have not done before. In this respect, it may open new trains of thought and discussions on the more difficult questions of existence itself.

WHAT IS TIME?

So how do we start such a book on what is seemingly an almost infinitely wide subject? Well, this book is an attempt to take a path that is mentally dynamic in all directions; it takes us through time based on cyclical recurrences from the start of time and ends who knows where, perhaps where time ends, if it ever does. This, being the start, may at first seem nonsensical, but it conveys the complexity of contemplating the state of being. Time is the key to how we perceive the things around us because it provides a frame of reference for what is a very dynamic existence full of oscillations.

The first thing to contemplate is time itself without which frequencies of any sort could not exist. What is time? In a lifeless void that is infinite in its dimensions, with no specific point of reference, would time exist? As self-aware humans, we need points of reference with recognisable starts and ends, in other words following predictable cycles and parameters. Looking up into the void of space and trying to conceive what it is in terms of its infinite dimensions requires a combination of time and distance, and to many, if not all, this is beyond actual mental comprehension. The thought that we could possibly take off from earth and follow a straight line into space and that the line would carry on forever, never ever ending, is a difficult concept and outside the realm of our experience. The question that could be put to anyone suggesting that there are finite limits to space, or the void in which the universe exists, could be, 'What lies beyond those limits?' Of course, the answers to this question could go on and on ad infinitum because if a barrier was ever discovered, or further ones, then logically some void must exist beyond all of them. What we see and experience in this existence is exclusive to us since our perspective on the universe is different from that of all others.

AN ANALOGY OF A RAINBOW.

Everyone able to view rainbow formations will see the rainbow from a slightly different angle than others. Even each eye has its own view and angle. When viewing a rainbow, the viewing eye or eyes are each exactly lined up with the centre of the rainbow in a direct line perpendicular to the rainbow's centre, as if it formed a complete circle, and moving through the eye to the centre of the sun. Rainbows formed in the mists of clouds, when seen from aircraft (see Figure 1.1), can be viewed as complete circles, with the aircraft's shadow seen in its centre and the viewer in a position within the aircraft's shadow in direct alignment with the sun. This makes each person's view exclusive to them.

Figure 1.1 Rainbow formed on a cloud when viewed from an aircraft. The aircraft's shadow is in the exact centre of the completely circular formation. (My photo.)

Such exclusivity means that each one of us sees a rainbow that is different and not seen by any others. Individual exclusivity can also be applied to the infinite void in which our universe exists. Since it extends ad infinitum equally in all directions, then every point within such infinite dimensions could be claimed to be the exact centre. We can all claim to be at the centre of the void containing our universe and our very existence, although not being at the centre of our universe per se.

INFINITY

The term *ad infinitum*, meaning forever, is now in our minds where time and distance must logically be related from a starting reference point from where we are on earth. Since there cannot be an actual start or end point of that imaginary line in an infinite void, it is more difficult to visualise as there are no referential frames that could possibly terminate it. So, the concept of an infinite physical distance becomes intertwined with a further concept of a never-ending time frame. However, time would appear to be a one-way street in that it also originates in a past infinity without a specific starting point. This would mean that time has always been running, but what we individually experience is a minute segment of that time. This, however, unlike omnidirectional spatial infinity, cannot be reversed. The term *omnidirectional spatial infinity* can be further visualised in that our imaginary line is just one of an infinite number of possible lines, each radiating from our starting point three-dimensionally, that is in all directions with the same never-ending results, meaning that there is an infinity of directional infinities. If this is also a difficult concept, then take the first imaginary line and imagine it coming from infinity and going through the centre of the earth and carrying on to infinity in the opposite direction. This is shown in Figure 1.2.

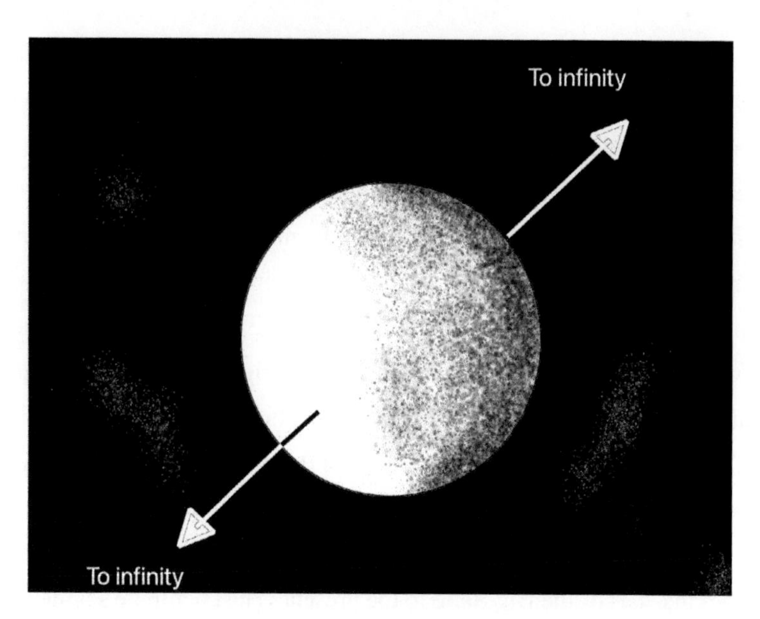

Figure 1.2 To and from infinity × 2.

Since in one direction the line comes from, or goes to, infinity, and in the other direction it does the same, then we must have two times infinity. Repeat this from every perspective on the earth, every star, and every other planet orbiting every other star in our, and possibly every other, universe, or individual points throughout, etc., the angles of which are also an infinite number in themselves. We therefore end up with an infinite number of directional infinities. In other words, the void in which the universe exists is infinite in all three dimensions.

This sets the basis for our existence in that we must take for granted certain facts about this existence that can never be proven, that is, in this infinite void containing our universe, the point from which we specifically originated and from which reference in time it could have all started. However, everything in our perceived existence may have one thing in common, a frequency component, and by using this is how we can theorise how energy began with oscillations of all frequencies that affect us and everything around, which is what provided the motivation for this book. It would be an impossible task to cover every one of the frequencies of existence in sufficient depth to make this book a complete body of reference and an information source for every cyclical event in existence. This was not my intention. Rather, the intention was to stimulate interest for those readers who may not have a leaning towards studying physics, chemistry, or general science in any great depth. I will be making general assumptions about a starting time frame, a point in the past from which all things that we currently know of originated. This is required for the sake of measurements with regard to space, time, and distance.

Of the many theories put forward, the Big Bang theory is the most plausible. All measurements and frequencies will be using this estimated starting point as the beginning of time in our universe as we know it. This notwithstanding that from the start of the Big Bang to the present is just a minute segment of an infinite time period.

PERCEPTIONS AND BELIEFS

Our senses are tuned to take in information originating from a variety of sources. The most obvious of these sources are light, heat, touch, smell, and sound sensations, all of which are oscillatory, having a frequency component from external stimulation of bodily sensors that generates internally in the form of direct interactions with the body. All these senses, which eventually generate varying frequency impulses in the form of signals to the brain from external stimuli, are just examples showing there are frequencies interacting with all aspects of life. A single nerve impulse, called an action potential (AP), would have little or no effect. It is a sustained chain of APs that is required to stimulate various centres of the brain to elicit a reaction, and all these have frequency rates dependent upon the intensity of the stimulation. However, there are other external sources, not immediately apparent, that interact in some way not just with us but also with virtually every living thing and object around us forming the very basis of existence, gravity being a prime example. Some of these will be included in our discussions as the book progresses.

Perception of our surroundings and our survival requires that we detect energy that is dynamic i.e. that is it has frequency. In space, the origin of this dynamism is derived from many sources but has a constant speed of transmission and is, for the most part, electromagnetic in nature. On the earth we have evolved to directly interact with many different wavelengths of radiative electromagnetic energy, such as light and heat, amongst others, along with other oscillations such as sound waves that are mechanical in nature. Because we, along with all other sentient and non-sentient beings, are born, grow up, and react to all inputs and sources of radiative energy, we therefore become products of our environment. The sensory inputs detecting dynamic energy to enable our awareness of that energy are in the most cases apparent, but we rarely question their natural origins.

It may be true to state that most people in life do not contemplate their own existence or the physical origins of objects that they daily encounter, let alone the origins of life. How many people look at a rock or stone and think about just what it is made of, where it came from, and how and why it was formed, let alone the dynamics of its atomic structure? Also, how many others really take stock of the fact that we live on a giant sphere made up of those rocks along with other elements, including metals and water, and a myriad of chemicals? This collectively surrounds an incredibly hot core floating weightlessly in a vacuum called space, all this whilst orbiting a massive source of energy called the sun. In addition, everything, including the moon and other planets, is held in place by a mysterious universally attractive force called gravity, which extends its influence far beyond everything that is visible in the outer reaches of our universe.

The Latin phrase *cogito ergo sum*, meaning ‘I think, therefore I am,’ may unconsciously sum up an acceptance of being, and perhaps for most people working and living in their own safe environment and situations, this is probably the norm. Many never really ever feel the need to question their own existence or its deeper origins. However, throughout ancient history there may have been certain people who, when looking up at the stars and planets, perhaps started to question and then to theorise as to how we and everything around us came into this existence. Different interpretations based upon the limited knowledge available at various times throughout the ages may have given rise to the populations of those times listening to, and accepting, these ideas, which would then have been passed on by word of mouth and added to over time. This may have further resulted in establishing those original thinkers as wise men and women, prophets and sages, whose followers began to form belief systems based on their thoughts and deliberations. These, in turn, may have formed the basis of, and eventually evolved into, the many forms of religion.

In those ancient times, when people perhaps followed a subsistence existence, it may have been much simpler to accept and say that a god, or gods, created everything without having to consider the how's and whys, because it takes the burden of deep contemplation and analysis away from having to question their own existence, especially if someone has done it before and it has developed to become the accepted norm. It also probably made for a more peaceful life as authority and dogma became entrenched in primitive society, where questioning those beliefs would have been seen as a threat to the then established order of things. Typical of such dogma are the simplistic explanations found as a common theme in many religions but typified in the biblical book of Genesis. Here we find that an undefinable entity, in terms of origin or form, called God created each element of the universe at his mere command. This is followed with similar explanations for the creation of the earth, the sun, people, animals, plants, and so forth. These beliefs have held sway for millennia, and many, probably most of the world's population, still believe them in some form or other. The 'popping' into existence, as I like to call it, may be true in part given the current thinking about the Big Bang, but it is evidenced in a reasoned and, to a certain extent, calculable way. However, the formation of all that we know of exists in both living and non-living form through energetic dynamism arising from that Big Bang. It is my belief that the substance of such dynamics, including the formation of all matter, can be broken down into oscillations that define each aspect of existence.

Not until early classical Greek times did the foundations of modern science start to evolve and challenge established thinking, for example the ancient Greeks' discovery of the heliocentric system (Petrakis 2004). This took another few thousand years to arrive at a point where free thinking and expression was allowed and the prevailing religious dogma was able to be challenged without repercussions, although even today in some cultures, to challenge specific religions and their teachings can carry severe penalties. In my own experience, I have looked at religion, and once even played a guitar for a small Christian gospel group, but underlying this I always questioned statements taken as fact by those around me.

LOOKING AT THE MOON

One of the more obvious ones that always stood out for me from a very early age was that God created the sun to light the earth by day and the moon by night. As a child in junior school, I would often get into trouble with the teacher for not paying attention because I was staring out of the window at certain times when the daytime moon was visible from

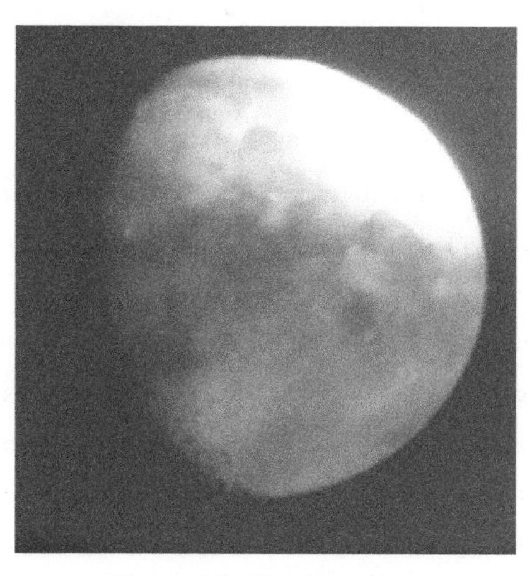

Figure 1.3 Daytime moon.

my desk when this was positioned close to a window.

Even at that very early age I was fascinated with this when most children my age either simply ignored it or took it for granted. I was even more entranced when I realised, and could see, that the moon was more than an ever-changing crescent shape but a three-dimensional sphere. This three-dimensional realisation of what I could see came suddenly. It instantly became to me 'another world' that did not provide the source of light but simply reflected sunlight. It also seemed to me, in my childlike way, that in the new moon's waxing phase, the bulk of the reflected light formed a crescent like a bow with an imaginary arrow across its centre that aimed directly at the sun (Figure 1.4).

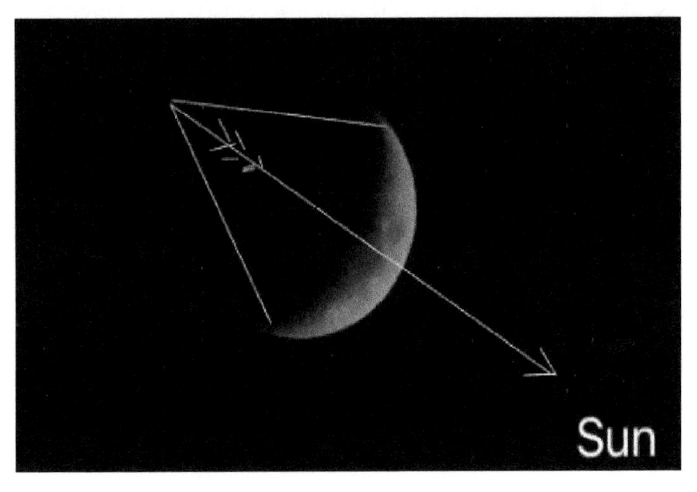

Figure 1.4 Orientation of the moon to the sun in the early phase.

As the phases of the moon progressed in the waning phase, still at that very young age I realised that the orientation between the light and dark areas was always in a line with the sun. Some questions even at that time began to form in my mind about things seemingly accepted, but never questioned, where certain religious statements are taken as facts. My religious education teacher had quoted a simplified form of the passage found in Genesis 1:16: 'God made the two great lights, the greater light to govern the day, and the lesser light to govern the night; He made the stars also.'

True, I thought. *At certain times the moon does shine brightly by night, but I see it just as much at other times in the clear blue daytime sky.* See Figure 1.3. This is usually followed when a clear night sky is totally dark except for the stars. This conflict of religious dogma and observable facts made me start to question such statements. Even at that young age I reasoned that the moon would always have to be on the opposite side of the earth from the sun in order to only ever be visible at night, whereas in reality, the only time that the moon is exactly on the opposite side of the earth from the sun, that is in the dark, is during a full moon phase, especially culminating when a lunar eclipse takes place. See Figure 1.5.

Figure 1.5 Lunar eclipse.

This is where the moon falls within the shadow cast by the earth. Sometimes this makes the moon seem red, a 'blood moon'. The reason why the moon turns a reddish colour is explained in Chapter 11. It is the moon's orbit around the earth providing different angular orientation towards the sun that gives rise to the waxing and waning phases, showing as crescents of dark or light areas both in the daytime and night-time skies, depending on where on earth the view is seen from. These early years of mine led me later to think of other lunar repetitions that are very predictable. These will be further discussed in Chapter 11.

ETERNITY AND INFINITY

Over the years, other forms of repetition have intrigued me, starting with the Big Bang theory and the convincing evidence that it did occur and that it could itself be part of a grander cycle of existence that stretches backwards and forward for eternity.

Eternity and infinity are expressions of something that is never ending, both stretching forward and backwards from and through time. Eternity specifically adds a time dimension, whereas infinity is related to physical size and distance. All existence is only observable and sensed through various forms of oscillations, which I refer to as 'ripples in the ether', originating from that initial starting point for this particular cycle of existence. Then, by looking at the many types of repetitions, henceforth referred to as 'frequencies', this caused me to contemplate and gave rise to the truth that the dynamism and nature of all existence follows the same pattern. My aim is to look at these, from extremely high frequencies found within the smallest subatomic particles, progressing to the frequencies of gamma and X-rays, and continuing to work our way through the range to the very low frequencies and discuss where they fit in with life, health, and developing technology as we understand it.

SCIENCE AND PERSONAL THEORIES

As this book is intended to be a personal excursion into the dynamics of existence, then within its pages you will find personal ideas and assumptions based upon my own thoughts and not necessarily referencing the work of others. However, some of the important milestones will still be referenced. Much of the book is derived from personally applied logic backed up by experience from my studies of electronics, physics, and environmental chemistry leading to medical research. So, it is not my intention to make this a deeply scientific or medical therapy book, but some scientific theories and medical uses will be discussed in the text, along with simpler calculations where needed. I also include how various frequencies may be put to work and try to give some explanation of the technicalities and the interactions with uses in everyday life, particularly in the field of communications.

It is also not my intention to criticise those whose acceptance of reality means that they prefer to choose explanations based upon religion. However, I hope that since I, and far greater minds than mine, have put forward ideas that would make the whole universe appear to be governed by rules that could be compared to those from a living pulsating deity, then there may be some common ground between religion and science. I would call these the general, but not absolute, laws of the physical universe, and to me they are as wondrous as any religious interpretation. As to how these laws came into being is a greater unknown and perhaps beyond any scientific concept currently around or ever likely to be understood. Maybe this is from a collective intelligence beyond our comprehension and may seem 'godlike', but in my opinion it is far greater than that simplistic term could ever portray. I have heard many religious people argue that that the randomness of creation arising from the Big Bang could never have led to the complexity of DNA needed for life to occur, and therefore life must have had some form of intelligent design. But those same people happily accept that an infinitely more complex entity somehow came into existence. I suppose the vague counter to that argument, in my opinion, is that in an infinite void all things are possible, but that gravity and energy and their frequencies are the central basis of all possible things.

As an individual I have always tried to be logical in my approach to most things that are both emotional and physical, not always with success, but in an overall positive way. However, there are certain rules to these physical laws that do not conform to logic. The most obvious of these is related to the speed of light. I have tried to reason around this ultimate speed anomaly, and in my discussion, I put forward my own theories that may or may not tie in with current scientific thinking. The great thing about being an individual is that one's unique ability is to think about an issue and then theorise. Whether one's theories are correct is fully open to debate, but the theories can start discussions amongst other serious thinkers that may lead, eventually, to the truth.

Chapter 2

THE RESONANCES OF EXISTENCE—IN THE BEGINNING

A book about frequencies covering the whole spectrum would be very narrow in content and very boring if just the numbers were used and listed. If we intertwine those frequencies with various factors, including discussions on cause and effect, then the numbers take on a meaning that can be related to. For the non-scientist or non-mathematician, some of the numbers discussed may seem beyond comprehension, both because they are very large and, conversely, extremely small. However, trying to give an understanding to existence requires that all aspects be included with the fact that the basis of virtually everything around, whether solid, liquid, gaseous, or plasmatic, can be broken down to oscillations, with these encompassing all frequencies found throughout and at the extremes of very wide spectrums. This chapter attempts to start off this process.

THE MUSIC OF THE SPHERES

A phrase often referred to from antiquity that has always intrigued me in both an artistic and scientific way is 'the music of the spheres' (*música universalis*), which is said to be an idea from Pythagoras (569–490 BC): 'There is geometry in the humming of the strings, there is music in the spacing of the spheres' (Young 1965). This phrase has its origins in ancient Greece, where a belief arising from observations of the planets was that the celestial bodies moved rhythmically at different rates that had a sort of harmony, not in a literal sonic sense but with the frequency and harmonics on a different scale, one requiring some imagination. Johannes Kepler (1571–1630) said the following about the music of the spheres: 'The heavenly motions are nothing but a continuous song for several voices, perceived not by the ear but by the intellect, a figured music which sets landmarks in the immeasurable flow of time' Kepler provided the mathematics for planetary orbits around the sun that remain accurate to this day.

It is likely that when the Greeks were referring to motional observations of planets, they assumed that the planets were spherical, this whilst observing them in relation to the background stars. This observation is from a moving platform called the earth. The observation gives an ever-changing perspective in which the planets appear to have gyrations that, when viewed from the earth, move rhythmically as the earth moves faster in its orbit of the sun than those planets do in their own orbits. See Figure 2.1.

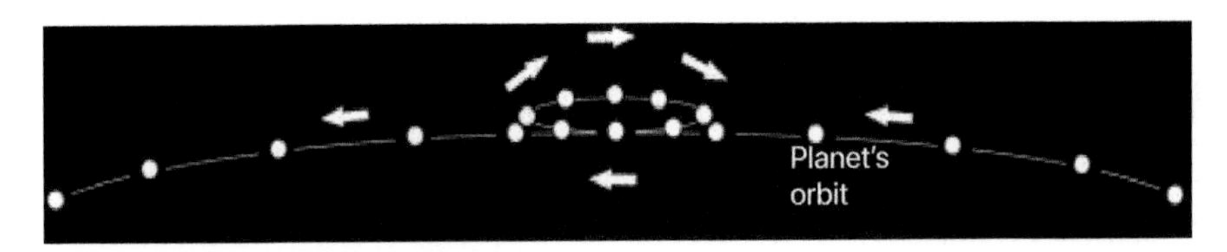

Figure 2.1 Slightly angular aspect of planetary motion viewed from earth.

MEASUREMENTS

These dynamic visual qualities, and others, on such a vast scale have frequency, and it is this that is intriguing since it is natural to think of frequency in cycles of events where the cycle repeats itself after a specific time period of hertz (Hz), which is a unit named after physicist Heinrich Hertz (1857–1894), whose research into electromagnetic waves led to the ability to calculate their wavelength measurements and is used as a relative timescale to the number of repetitions per second, or in old terminology, cycles per second. In order to standardise terms and provide direct comparisons, a single second will be used as a general reference, along with wavelengths—in other words, the time taken when measured from peak to peak of a single cycle and the number of those 'cycles' of events that occur in one second. This allows the use of the term *hertz* (Hz) as a common referenced time period throughout the book for both the very high frequencies and those with cyclical time periods in the billions of years. Figure 2.2 shows, graphically, two simple cycles in relation to time.

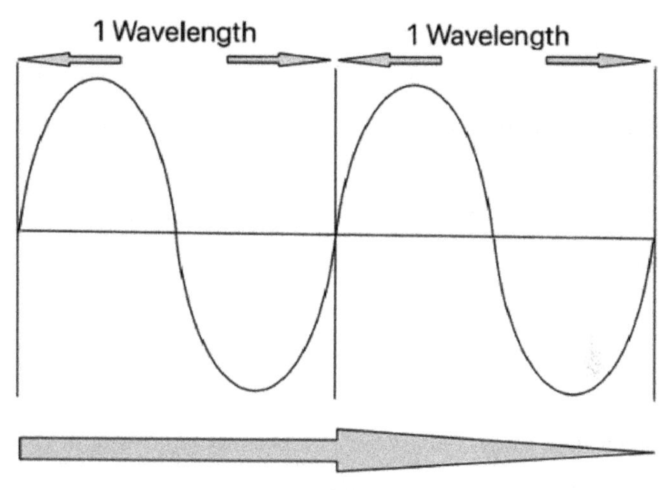

Figure 2.2 Relative time.
This represents two cycles of an event. If the relative time shown equals one second, then the repetitive frequency equals 2 Hz. Conversely, if the relative time was four seconds, the frequency would equal 0.5 Hz.

To put this in context, a yearly occurring event such as the instant the sun rises during the summer solstice at Stonehenge would occur about once every 365.256 × 24 × 60 × 60 second, equalling the number of seconds in a year. This is calculated by taking days multiplied by hours, multiplied by minutes, then multiplied by seconds, giving a value of 31 558 118.4 seconds. To give a value in hertz would simply be to say one second's worth is the ordinal 1/31 558 118.4th of the total number in a year.

For the nonmathematically inclined reader, to simplify this in terms of hertz, dividing the number of seconds in a year into 1 gives:

$$1/31\,558\,118.4 = 0.317\,10^{-7}\text{ Hz or } 3.17 \times 10^{-8}\text{ Hz}$$

(Figure rounded up to two decimal places)

This also provides a frequency rate of the earth's complete orbit around the sun. Because this sort of value is below 1 Hz (less than one cycle per second) I have called it, and others to be discussed later in the book, a 'subunity' (less than 1) frequency value.

ELECTROMAGNETIC ENERGY

The term *ether* in the title of this book refers to the now discredited belief that light and radio waves needed to travel through a medium in much the same way as sound propagates as pressure waves through air. In the nineteenth century it was thought of as a physical universal substance acting as a medium for the transmission of electromagnetic radiation, including all that will be discussed in this chapter. It was assumed to be weightless and transparent. This will be further discussed in Chapter 10, but in summary, sound radiates by causing compression and rarefaction of air molecules as the sonic energy radiates outwards from a source of the sound, thus causing the vibrations or ripples in the surrounding air.

A two-dimensional analogy makes it much like the ripples in a pond from a stone tossed in, but in the case of sound, this occurs naturally in three dimensions—but it requires the medium of the air or solids in order to be able to propagate. The special nature of electromagnetic waves of any sort means they do not need an intervening medium or 'ether', and all will travel at a constant speed, the speed of light, through a vacuum. However, use of the term *ether* in the context of electromagnetism persists, but now in a more esoteric way to describe the intervening space, the vacuum or void through which the emitted or transmitted electromagnetic energy transits.

Electromagnetic energy will later be referred to as photons or electromagnetic radiation. The difference between being 'emitted' and 'transmitted' is that photons are directly 'emitted' from atoms (the mechanisms of this will be later discussed). Transmitted electromagnetic energy usually refers to the energy 'radiated' from antennas but travelling at the same light speed thus sharing some of the characteristics of photons. Both emitted photons and transmitted electromagnetic energy are forms of radiation. These includes human derived transmissions to and from mobile phones, televisions, radios, radar devices, walkie-talkies, etc., as well as those used in satellite communications, and will also be discussed in later chapters. A medium is certainly required for the transmission of other mechanical vibrations at the lower frequencies, as discussed above, through gases, liquids, and solids, including the earth itself, but the term *ether* was really only applied to electromagnetic waves. Later in this chapter a theoretical discussion about the limiting factor of the speed of light will suggest that there could be a hitherto unknown physical property in space that may be responsible for restricting light's speed through it.

Imagine that an infinite void exists where there is no energy. The void would be an absolute vacuum that has no discernible dimensions, temperature reference, or time, but more importantly, no beginning or end to give any such reference points. Did such a situation ever exist? This was discussed to some extent in Chapter one. Astrophysicists, physical scientists, and astronomers would probably say that the foregoing is a pretty good description of what it may have been like before the Big Bang; in other words, there was an absolute nothingness. This is a concept difficult for humans to perceive since our experience has references to starts and finishes, decay and renewal, birth and death, with always something to follow or to repeat. However, in the matter of life and death, what follows is speculative and based more on hope and faith than on absolute provable fact, but even then, with energy and mass conversions—and as we are all a manifestation of energy—there is a continuation in some form afterwards even if just as totally disordered energy, also known as entropy. The term *entropy* has many definitions but is, in essence, where energy goes from an ordered useful form to a disordered form in the course of being transduced into other, less directed forms.

CREATION OF THE UNIVERSE

The universe as we know it, and according to the Big Bang theory, came into existence from a single point, a singularity, where all subatomic and energetic particles exploded into existence and were formed well within an immeasurably short amount of time. This period of elapsed time has to be put into context because time did not exist in the much slower passage of it than what we would now perceive. That rapid expansion formed the basis of all the particles that gave rise to the substance of existence and also gave rise to time itself. This came about because the instant of the start of the Big Bang gave the first universal timing reference point for everything that followed in relation to it.

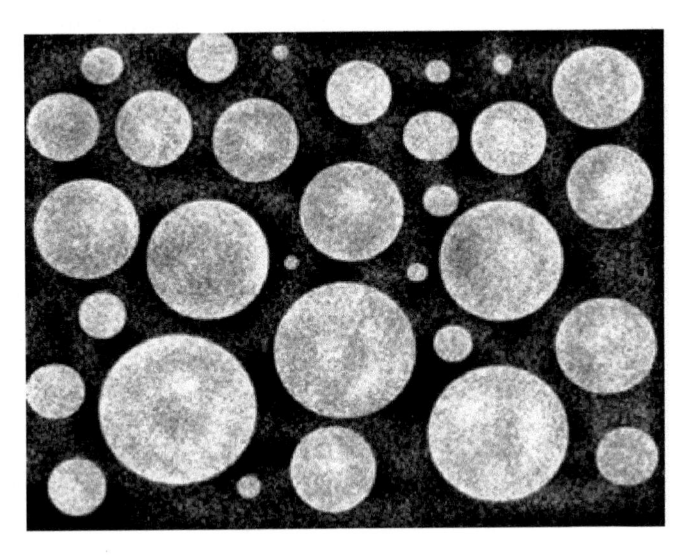

Figure 2.3 Visualisation of globular universes stretching to infinity, all possibly expanding and collapsing.

However, there is a fly in the ointment of the Big Bang theory in that some distant celestial object recently discovered by the James Webb Space Telescope, along with others, may predate the Big Bang (Sutter 2020). This would suggest to me that Big Bangs are not unique and confined to the one we believe to be the start of our existence, because that which has been recently observed may be large stars or galaxies at the outer reaches of other universes from other Big Bangs. See my following discussions and Figure 2.3.

ABOUT THE CHRONON

We divide time into years, days, minutes, and seconds, but the minimum theoretical time that can occur between two events, especially at the subatomic level with events happening sub-atomically, gave rise to what is called the 'chronon'. The chronon is a theoretical time-derived quantity related to the time it takes light to travel across the classical radius of an electron.

In quantum physics the chronon suggests a link between time and gravity and is indivisible. It was derived from a proposal by Henry Margenau, a German American physicist, in 1950 for a specific time measurement. It was suggested from calculations by Piero Caldirola to be around 6.27×10^{-24} seconds in duration and is stated to be the lowest indivisible unit value of time. However, another great scientist, the German physicist Max Planck, calculated a much smaller time frame of 5.39×10^{-44} seconds, a value known as 'Planck time' or a 'Planck second'. Again, it is a measure using the speed of light to transit time over a specific distance. In this case it is a Planck length. Without going into the deep physics and dimensional analysis, I would summarise that it is a value derived from the mathematical combination of a number of different constants that include gravity, the speed of light, Boltzmann's constant, and Planck's own constant. It is said to be a natural value not derived from human-related dimensions such as metres or feet, yards or miles, and is sometimes called a 'God unit'.

The original definition of a chronon is still a hypothetical quantity. The foregoing original time value in relation to the possible physical dimensions of an electron will be added to later in chapter 3. Also, gravitons that give rise to gravity are a theoretical possible quality derived from the very specific high frequencies formed in the instant following the Big Bang and are probably the most essential part of creation. These will also be discussed later in this chapter and also in Chapter 11, but know for now that they have a direct relationship to the speed of light. These frequencies carried the energy that then combined to form all the elements that we currently know of and affect such things as time, gravity, and mass.

BIG BANG AND OTHER THEORIES

The energy involved in the Big Bang is possibly beyond human comprehension, and as it theoretically came from a singularity then, to my mind, therein lies an enigma. In physics, the first law of thermodynamics states that energy only ever changes or is transduced but is never created or destroyed. The point from which everything came, the singularity, must have existed in some form before the almighty explosion took place. If this is true, then looking at the lowest possible frequency of existence gives us a value that is directly related to the formation and subsequent expansion of the universe, but as part of a repeating progression.

This fits in with another theory, the oscillating universe theory (Saslaw 1991). This oscillating universe theory, although currently unpopular in some scientific circles, with scientists favouring the theory of inflation (Guth and Steinhardt 1984), is one that allows us to begin to make some calculations of the rate of repetition based upon these theoretical expansions and collapses. The idea of a singularity producing everything in the known universe and coming from a single point is a difficult concept to many. It has no definable dimensions, and its origins are unknown. To me it would appear to be the opposite end of the infinite-sized dimensions of the void in which the universe exists, infinitely small compared to infinitely large, and beyond visual comprehension in both respects.

The oscillating universe theory suggests a Big Bang that causes the universe to expand but is limited in its outward expansion and will eventually contract under the forces of gravity (also created by the Big Bang) to bring everything with ever-increasing rapidity back to a point singularity called the 'Big Crunch' (Turok 2004). The compressed energy would once again be so huge that it then would explode again and, in so doing, produce another Big Bang. (This theoretical series of events will be added to and further discussed below.) This process would repeat itself ad infinitum. The oscillating universe theory has many critics, but if this theory is correct, then the frequency of these oscillations counts as being the lowest one possible for us to attempt to calculate. The actual rate of these oscillations is unknown, but the current estimate of time elapsed since the Big Bang is around 13.7 billion to 13.8 billion years, so we can guess a figure of well above 30 billion years to complete a cycle of expansion and collapse, although this is pure supposition. A value based on one 30-billion-year complete oscillating universe cycle will be given later in this chapter.

Our observation of the universe is limited by the speed of light, and that is directly coupled with the time that has elapsed since the Big Bang. The universe is most likely expanding globe-like in all directions and at high speed, in what is called 'isotropic expansion'. However, the maximum rate of expansion applies only to photons emitted at the instant of the Big Bang, these travelling at the speed of light. To put this into context, a traveller from outside the realms of our universe but travelling directly towards it would, in the first instance, be aware of its existence when encountering those first photons emitted from the Big Bang. These would be the first photons of electromagnetic radiation ever created and emitted directly from the explosion. This would then constitute the still expanding outer boundaries of the 'photonic' universe at the instance of first encounter. With the expansion of physical matter, observational evidence of the physical universe using red shift analysis (see below) confirms that galaxies at the most distant reaches of the universe are still in the expansion phase at speeds lower than that of light and also appear to be expanding in an isotropic manner, outwards from a single universal point. This is suggested to be the point of origin of the original singularity, which gives extra credence to the Big Bang theory.

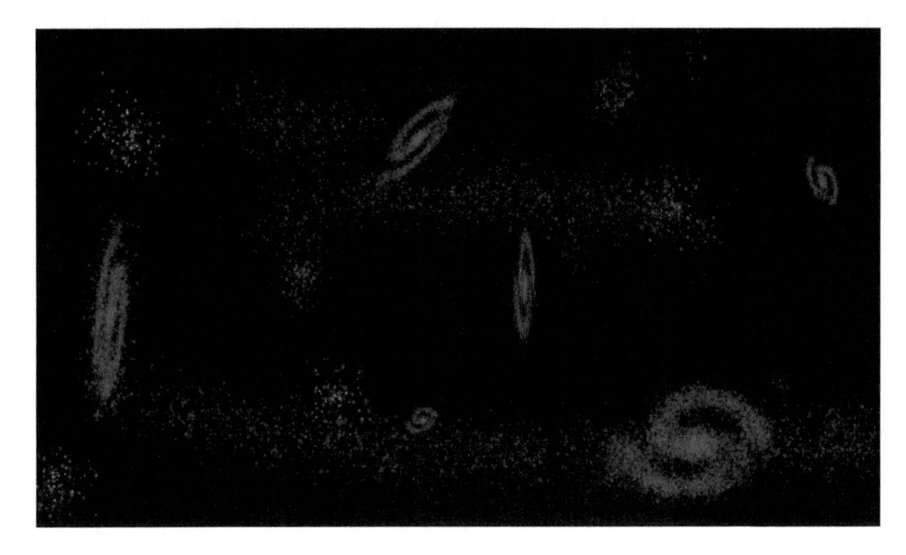

Figure 2.4 Visualisation of the most distant galaxies receding as the universe expands. All appear red.

NB. Red shift is the phenomenon of the observable change of colour of light emitted towards the earth from objects receding at high speed away from an earthbound observer because of the Doppler effect, extending the wavelengths towards the red end of the visible spectrum. This is the same as the sound of a whistle which appears to change its pitch for the observer as its source travels to and from a stationary point. This red shift occurs only with light emitted towards an observer from an object moving away. If an object were approaching and emitting light, then there would be a blue shift. The speed of light remains constant in all cases. A discussion on aspects of this is found in Chapter 4.

Using the oscillating universe theory to give us some form of cyclical measurement, we see that one complete cycle from the Big Bang to the Big Crunch will be greater than 2×13.8 billion years (my estimate above suggests 30 billion as a rough value). This assumes that the estimated time from the Big Bang to its maximum expansion to that of the Big Crunch of the universe is equal.

The energy expanded in the Big Bang explosion would be at its maximum at the instant of the explosion. This energy would then become kinetic, reducing in value through time as it follows the inverse square law until fully depleted but, in the process, forming all the galaxies and stars as 'gravitational eddies' start to develop amidst that early expansion. My use of the term *gravitational eddies* means that uneven gravitational anomalies would start to attract particles together in spinning clumps. These would eventually form a higher gravitational anomaly, increasing in mass towards the centre of what would initially become a spiral formation with a solidly forming centre as a result of the increasing gravity. These masses would eventually form as stars, triggering nuclear fusion reactions within them through strong gravitational forces.

The more massive of the stars may then, under extreme gravity, collapse in on themselves as a supernova. The crab nebula is an example of the remnant of a supernova. See Fig. 2.5

Some supernova from the more massive stars form black holes. This then creates the conditions for attraction of other stars to spiral towards a central part of the eddy, which eventually gives rise to galaxies circling and spiralling into 'black holes', and there are many billions of galaxies in this still expanding universe's outward growth.

At some point gravity would, in theory, fully take over and start attracting all the formations in the universe back together, including the entropy resulting from expended energy. This would slowly increase the density and energy

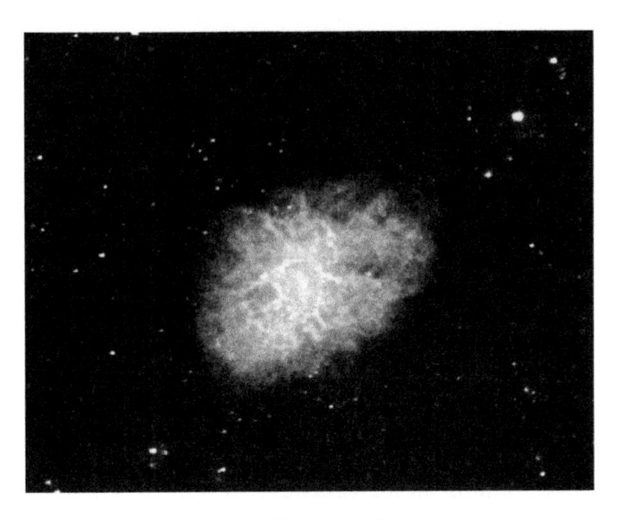

Fig 2.5. The Crab Nebula
©2022 Robin Barrett Astro-photographer
Remnant of supernova with filaments extending from the central pulsar.

until all particles and energies that make up the universe are attracted back into an enormous gravitational black hole caused by the formation of an ever-increasing tightly condensed mass. The gravitational force would then become increasingly higher so that it would eventually compress all matter, once again making it a singularity, previously referred to in this process as the Big Crunch. All the energy in the universe would then be recaptured and be sufficient to cause another Big Bang. This is hypothetical, but could arise as a result of the inertial energy of all the universal material accelerated to such a high speed, perhaps beyond the speed of light and the laws of physics, as it comes together such that it carries on beyond the final instant of the Big Crunch, creating another Big Bang. This could be similar to a supernova in that the new Big Bang occurs when the outward thermal pressure exceeds the gravitational pull of the central core.

Of course, beyond the void in which our universe formed, there exists the possibility that there are other universes formed and still forming from their own Big Bangs. These thoughts allow the mind to imagine that all the universes are possibly expanding and collapsing in globular form similarly to, and from, single-point singularities. Perhaps these universe 'bubbles' are spread throughout the infinity of the void, or in clusters, rising and falling as they oscillate from Big Bang to Big Crunch, causing another Big Bang and so on.

This may further allow the mind to visualise from an observational point of view, in space and time and measured in billions of billions of years or greater, how that vast region would appear. It may be like a seething cauldron of energy spread throughout existence, bubbling away in three dimensions. It also allows for another hypothetical possibility, namely that there could be overlaps between the universe's 'bubbles', affecting one another with the possibility of one universe's 'bubbles' growing in mass and energy at the expense of another. Earlier in this chapter, I suggested that bodies recently discovered at a distance great than the current dimensions of our universe could be part of another universe. This extra universe theory could possibly fit in with this overlap idea. See figure 2.3 shown earlier in this chapter.

Another simple analogy of the expansion and subsequent collapse would be to think of gravity as a sort of unbreakable elastic string attached at one end to the original singularity. As the universe expands in every direction, every particle in the universe is explosively pushed away from the singularity, travelling for billions of years, until the stringlike elastic reaches its limit. Then it starts to retract all the particles, in all forms, to the original singularity, including all energy, again taking billions of years to achieve the Big Crunch. If the oscillating universe theory is correct, the whole process repeats and repeats and would, perhaps, have a resonant frequency.

THOUGHTS ON THE EFFECTS OF THE COLLAPSE OF THE UNIVERSE.

It may be fanciful, and almost in the realm of science fiction, to imagine that during the contraction phase a time will come, if the oscillating universe theory is correct, when all constructed matter then begins to deconstruct. This eventually would reduce everything to the fundamental particles that originally formed matter, almost as if time were reversing. This would be because the gravity gradient would increase exponentially towards the singularity as time progresses, stressing and breaking apart all mass. This is not to suggest that time as we understand it would run backwards, but all the structures, energy, and mass would eventually be broken up and return to the individual atoms by way of the increasing gravity gradient, and then to the subatomic particles and strings forming them.

In the early part of the contraction, the stars would get visibly larger in the sky and be more compressed in numbers as they were attracted towards and hurtling back to a common central point. This is in contrast with everything we now see expanding around us. All matter would be attracted back to that incredibly dense gravitational extreme that is at a single point of focus, or soon-to-be 'singularity'. The unique combinations of matter forming molecules and amino acids, giving rise to the DNA that gave us life, would eventually be gone and perhaps may not occur in the next expansion cycle or immediately in any others that may follow. Life as we know it may be a rare occurrence and unique in its form to the here and now as part of this particular expansion cycle. But it also gives the possibility that life and intelligence in other forms may have already occurred many times over in the infinity of time that has preceded us. In infinity there are an infinite number of possibilities available.

My analogy of elastic string theory is a simpler view of the main string theory put forward by astronomical and particle physicists (Schwarz 2000). String theory suggests that each thing that exploded from each Big Bang is in some way attached to everything else, including energy and all particles. It is this expanding attachment that would eventually cause all matter and energy to collapse back to that original singularity, then forming a new one and hence creating the conditions for a new Big Bang to occur.

For the purposes of this book, it is assumed that a complete cycle from Big Bang back to Big Crunch takes in excess of 30 billion years. Using 30 billion as roughly the starting point, one earth year is just one 30 billionth (30×10^{-9}) of the cycle. See Figure 2.5. Since there are approximately 31 558 118.4 seconds in a year, the resonant frequency of the universe in hertz (cycles per second) would be"

$$30 \times 10^{-9} / 31\,558\,118.4 = 9.5 \times 10^{-16} \text{ Hz}$$

It may, of course, be a much longer time period for a repeated cycle but would still be in the same order as given. However, other theories must be considered that may or may not have a frequency of repetition. Figure 2.6 illustrates the oscillating universe theory showing repeating rises and collapses of the universe over vast periods of time. It may be that if the oscillating universe theory holds true, then each cycle may lose some energy into the void. This would mean that at some time in the far distant future there would be one last cycle, with insufficient energy to create another Big Bang.

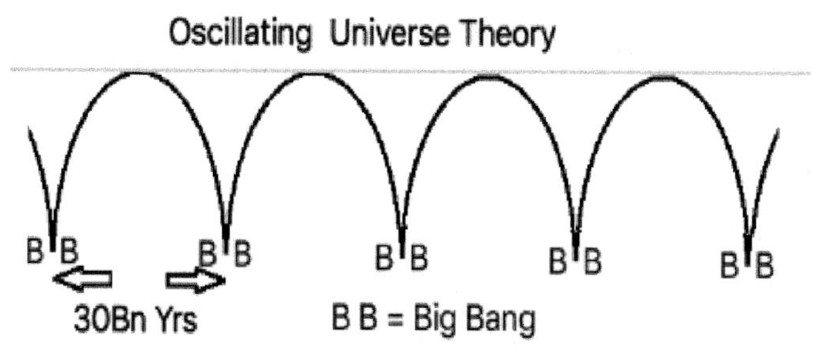

Figure 2.6 Oscillating universe theory rate.

If the universe is just expanding from the singularity and not oscillating in any sense, then there still may be a frequency at which these starting events take place but that is outside the realm of our universe and therefore not calculable. However, our current universe would carry on expanding until all stars and planets, along with their energy, would be very distant from each other and from the original Big Bang start point, with the energy and the effect of gravity being no more. What would be left would be cosmic electromagnetic dust on an incredibly vast scale. However, this leads to the possibility that there may be other singularities within the current volume of our universe that either randomly, or in a regular fashion, explode as other Big Bangs, as discussed above. These could have occurred within our time frame of 13.8 billion years, but the effects of which would still be in the realm of the unknown, beyond the range of viewable light, and because that light is coming from such distances, it has yet to reach any earthbound detectors. For example, supernovas that occurred in other galaxies millions of years ago are only just being seen in our current time. This raises all sorts of questions as to the eventual outcome and future of all matter formed.

INFLATION THEORY.

The foregoing discussion fits in to a certain extent with the increasingly popular and previously mentioned inflation theory (Guth and Steinhardt 1984) on the origins of the universe. This suggests that an immense form of energy exists outside the realms or dimensions of the observable universe, which has occasional 'troughs' in the waves of density that drop so low that part of the still incredible energy breaks free as a singularity, and this then instantly inflates, causing the Big Bang to occur. See Figure 2.7.

However, some interpretations of this theory suggest that this inflation is 'balloon like', where all galaxies start off small at the surface of the balloon and expand as the universe inflates. If this theory were true, it would mean that all galaxies are equidistant from the central point of the Big Bang, with empty space surrounded by a 'skin' formed by the galaxies. This does not conform to observable fact given that our own galaxy, the Milky Way, is hurtling towards Andromeda and the two will collide in the distant future.

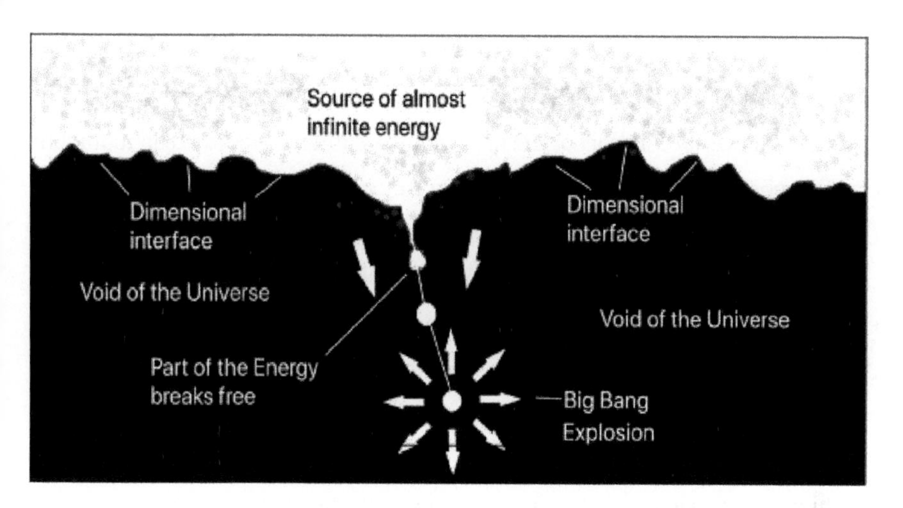

Figure 2.7 My illustration of inflation theory.

Inflation theory also gives rise to the idea that the frequencies of existence have origins that could also be common to other Big Bangs forming universes beyond our own observable universe, stretching into infinity. It also suggests to me that, if this theory is correct, we would be within a void surrounded by this energy, and therefore there would be a finite limit to the void in which our universe formed where there is a 'dimensional' interface with this energy. I would alternatively propose that the possible source of energy, following this theory, would have a finite size, forming a plasma-like ball of tremendous dimensions and be contained by gravitational forces. Occasional weakened gravitational anomalies allow a leakage into the surrounding infinite void such that the exuded energy, then still compressed by weakened gravity, forms the basis of the central point of the Big Bang as outward forces within the energy exceed the force of gravity. Perhaps this could be called 'the God ball' since it would be the basis of forming everything in the observable universe. This still does not address the origins of all this energy, which will forever be a mystery.

For our own observable universe, there is new research (Garner 2019) suggesting that the universe is expanding faster than first thought, but in some later calculations I have based the size of the universe on a maximum expansion rate limited by the speed of light if all the energy from the original Big Bang is included as part of the known universe. This would mean that there are two possible sizes for the universe, one being, as previously mentioned, the 'photonic' or electromagnetic universe, and the other being the physical universe. Both these volumes would be expanding, with the photonic universe expanding at the faster rate, limited only by the speed of light as previously discussed. The non-electromagnetic physical universe's expansion would be set by the explosive force of the original Big Bang. Red shift analysis would make it possible, with data from Earth-orbiting telescopes, to calculate the rate at which the physical universe is expanding and, by back projecting, the time elapsed since the Big Bang.

FREQUENCIES AND ENERGY

The theoretical lowest possible frequency in the universe is, as previously discussed, likely to be the oscillatory rate of the period from the Big Bang to the Big Crunch. All other frequencies created in the entire universe could be said to originate from the start of this, the lowest of low frequencies. However, as we progress, we will discuss some of the highest electromagnetic ones first, progressing to descending order. Some of the highest of these frequencies make up energetic photons such as gamma rays, but as we will see later, the incredibly high spin speeds and frequencies of subatomic particles exceed even these many times over.

The establishment of such high frequencies must have occurred in the first instance of that explosion as part of the expansion away from the singularity. Since the Big Bang, however caused, would have been the release of all energy in our universe, then there are questions that beg answers, such as 'What is energy, and how did it ultimately become solid matter?' The origin of all energy was discussed above, but the standard definition of *energy* in physics is 'the ability to do work'. Since work is always kinetically manifest as some form of motion, be it actual movement, or thermal agitation, or other oscillations, then I believe that the Big Bang represents the potential energy available, with kinetic energy, originating from the Big Bang, being the transduction of its release. Kinetic energy, in this context, could possibly be defined as 'the dynamic relationship between anything in motion relative to anything stationary or in a lesser state of motion'. It may be stated slightly differently by others, but this is my own interpretation. Kinetic energy in this way is interchangeable with potential energy depending upon how it is referenced.

Further discussion on what could be described as 'source potential energy' leads me to think of the original singularity as being like a compressed spring. My thoughts are that it is undefinable in its form and that it is difficult to comprehend what would have kept the energy confined within that singularity.

A more mysterious question is what would have triggered the release of all that pent-up energy if all confined energy remained in a singularity for some time before the Big Bang explosion. I believe that the oscillating universe theory may address this in that all energy is dynamic and cannot be confined. It goes from kinetic to potential (Big Crunch) and then explodes in an instant back to the Big Bang again as outwardly expanding kinetic energy. Original potential energy, in this sense, only exists in that instant between Big Crunch and Big Bang.

The Big Bang caused energy to radiate in all directions relative to its point of origin. As far as the potential of that energy's existing in a singularity, the only logical answer to its origins is that it has always been there in various forms of potential and kinetic from Big Bang to Big Bang, but given our limited ability to comprehend eternity and infinity, it is difficult to perceive. So, I would suggest that that which the universe is made of (being energy) is logically eternal and will continue to be so in one form or another forever, as discussed earlier in this chapter. This then links distance and time with energy in that all stretch from, and to, the same 'eternal infinity'.

Everything around us has a frequency. If it didn't, then it would cease to have substance since the frequency defines the mass. This concept gives rise to all sorts of theories, but a simple view here is that the frequency defines the substance. Albert Einstein provided a formula for the energy contained in a specific mass. Einstein's now 100-year-old formula $E = mc^2$ is still accepted today as the classical energy model where c^2 = the speed of light squared. This linking of the speed of light with mass and energy available would suggest that all matter must have an electromagnetic component and therefore a frequency equivalent when broken down to the basic elements making up that mass. However, working out an electromagnetic frequency equivalent for a given total mass is more difficult given the massive number of elements, electrons, neutrons, and protons in its make-up, all with different oscillatory characteristics.

In terms of electromagnetic energy, it would be possible to calculate the mass of an individual photon by combining this with another accepted formula by Max Planck (1858–1947) for the energy of any individual photon $e = h\upsilon$. From this we have $mc^2 = h\upsilon$ where h = Planck's constant and υ = the frequency of the photon. The formula then gives m (mass of an individual photon) = $h\upsilon / c^2$ and also shows that a photon's mass is also directly proportionate to its frequency.

Photons, packets of energy that make up most of radiated electromagnetic energy and defined by their frequencies. will be discussed in more detail in the next chapters. With mass in general, and since we are talking about potential energy available, then mass (m) in grams would equal the collective energy of all 'strings' vibrating at frequencies in all the elements and compounds making up the substance. The potential release of this energy in joules equates to $c^2 \times$ this value.

THE SPEED OF LIGHT

The speed of light c is equal to just under 300 000 000 metres per second (3×10^8) and is the limiting factor in the release of energy, such that the total potential energy is the speed of light squared multiplied by the mass in grams.

Ripples in the Ether II

At this subatomic level, it is possible that the laws of physics operate differently, and this may be due to the laws being created in the instant of existence until universal laws became established. From then on, the energy could only be released or radiated at the speed of light, currently believed to be the ultimate relative speed possible.

The speed of light is accepted as a universal limiting factor. However, there is rarely any discussion on why there is such a specific limit to light speed, with a seemingly universal scientific acceptance that it is the ultimate speed. It may be simply that individual packets of energy, called photons, are emitted at this speed given the combination of factors emitting them from atomic structures. There is also the possibility that a common threshold linking gravitational attraction may limit the release speed of photons. However, if we consider that the universe is not a complete vacuum as far as interatomic space is concerned, then we realise there may be speed restrictions to photons passing through it. If, at the moment of creation following the Big Bang, another, as yet undetectable material or force was also created along with the atoms and other subatomic entities that spread evenly throughout the expanding universe, this material or force may allow electromagnetic energy to pass through it relatively unimpeded. This may be in the form of a fine ether-like substance that has its own electromagnetic interactive rate, and it may be possible that we could recognise this rate as ultimately limiting the speed of photons.

Rationally, photons are bursts of electromagnetic energy that may also have a field formed around each and every individual one. Photons can be affected by gravity and slowed down below normal light speed when passing through wide crystalline structures such as glass. It may be that there is a form of intervening connected structure and that this gravitationally limits their speed of transit, possibly tying in with the 'string theory'. It may also be that the speed of light is universally restricted by this 'string' matrix if found throughout existence. This could lead to other theories, such as the idea that there may be regions outside our universe without this ether-like material where the laws of physics, as we understand them, are different.

Another possibility is that, based upon my theory, if some ways were found in our universe to exceed the speed of light, then there could be a rift created or a fracturing of that part of the universe that would also be outside the known laws of physics. However, this idea may fall apart in that when anything is transiting through any type of medium, there is usually friction or resistance in some form. This would eventually expend the energy by ultimately converting the motion into heat. In terms of their wavelengths, photons may lose energy and eventually become the lowest band of photonic electromagnetic energy in the far infrared. This will be discussed later in the book. This, at first, appears not to be so in the case of electromagnetic energy, because 'photons' of electromagnetic energy transit great distances over millions of years across the universe without any apparent measurable individual attenuation. However, it could be speculated that all electromagnetic photons will eventually somehow lose energy, possibly on account of some form of very minor frictional resistance or interference from other photons that effectively lowers their frequency. Inflation theory may suggest that this could be the ultimate decay of all electromagnetic energy.

This discussion is all the more interesting since dark matter is known to exist throughout the universe and may just be part of hitherto unknown matter, still waiting to be discovered, that came out of the Big Bang. Dark matter is thought to make up 85 per cent of matter in the known universe but is difficult to detect because of its inability to interact with electromagnetic energy such as photons. Particle physicists at the CERN accelerator are currently trying to identify methods of detection for this very elusive material. My suggestion about a matrix-like string structure and dark matter is that they may be closely related.

QUANTUM ENTANGLEMENT

Another theory, quantum entanglement (Freire 1970) described by Einstein in

1935 as 'spooky action at a distance' suggests that individual photons with a common origin can be split with the use of prisms and various devices and separated by great distances, but still maintaining an instantaneous link with each other. The action of one split photon at the quantum level is identically duplicated at the quantum level by the other, but in opposite response (up movement or rotation on one, down movement or opposite rotation on the other), caused by communication between them travelling at speeds many times (up to 10 000 times) the speed of light. Some sources suggest that speed does not come into it and that the response and communication between them is instantaneous regardless of distance, even if those distances are galaxies away. Because this defies the rule concerning the speed of light is the reason why Einstein called the phenomenon 'spooky action at a distance'.

Latest suggestions are that quantum entanglement is not restricted to just two split photons, but could apply to many billions of particles at the quantum level. Scientists still cannot explain quantum entanglement, but to me it is a manifestation of the interconnectivity of strings, stringlets, and gravity throughout existence where time, distance, and energy are closely related and may always remain an enigma.

Although beyond the scope of this book, I would simplistically describe quantum entanglement as being part of a quantum balancing act where the universe seeks to equalise or balance all energetic particles maintaining a sort of universal equilibrium, including split energy particles as with photons. This is theoretical because there is a link in the forming of a stringlike connection, especially where two particles arise from the same source. The speed of light, although limiting the speed of split particle separation, would not affect the balancing of the two entities since nothing is transiting between the two ends. It is just an unexplainable phenomenon of the conjoined nature of the two-split frequency-based entities maintaining a balance to their existence and being outside the established laws of physics. However, there may be a suggestion of this phenomenon being able to provide instant communication over great distances in space.

In my opinion, there is a catch in the usefulness of this phenomenon in that split photons can each only travel at light speed relatively in each direction. This would mean that initial communication over vast distances, although there is an instantaneous link, is maintained between them even as the distance between them increases at light speed. The split photon's transit would depend upon the time taken to separate the distance between them. This would mean that they will still be limited in establishing communication at a distance by the initial transit time of the speed of light, unless each split photon is dynamically contained at some distant point from the other. It could be further speculated that photons created, and somehow split, at the instant after the start of the Big Bang are now at the extremes of the universe. If true, quantum entanglement would suggest that an occurrence at one extreme of the universe is quantumly mirrored instantly at the opposite side of the universe.

Another enigma with splitting photons relates to their energy. The specific energy of an individual photo can be calculated by using Planck's equation. If a photon is split, then the energy of that photon would also be split. This means that either the frequency of each half is also halved or that there is an amplitude change to account for this. There does not appear, at this time, to be any scientific analysis or discussion on this.

As the potential energy locked within substances was formed within the first instant of existence, it may be that the energy, as we will see later, is not stored as normally diminishing kinetic energy at the quantum level in the peripheral spin speeds of the hadrons, electrons, and positrons that make up mass. This kinetic energy, as stated above, may defy the standard definitions in that potential energy is defined as being the ability to do work, with kinetic energy being work in progress, transducing or expending that initial potential energy. However, in the case of subatomic particle orbits and spins, the rotational speeds never seem to be diminished unless affected by other subatomic particles. If this spin energy is released, then the energy would be phenomenally high, creating a knock-on or cascade effect with other nuclei. Release of this spin energy is the basis of a nuclear reaction. A simpler explanation for potential and kinetic energy could be that all frequencies, spins, and orbits are the kinetic energy transduced from the potential energy being the Big Bang explosion itself. This itself gives an idea of the enormity of the Big Bang if such a large amount of energy was either slowly released, as in a nuclear reactor, or instantaneously released in a nuclear explosion of such a small amount of fissionable material.

ATOMS IN THE UNIVERSE

If the concept of everything in existence, including all energy, arising from singularity is difficult to imagine, then consider a few facts based on observations of all matter in the universe. There is a suggestion that the total number of atoms in the entire universe is between 10^{78} and 10^{82} (Space and Astronomy news – Universe today). For the purposes of calculation, we will use the given value between the two, this being 10^{80}. This means that the value 10^{80} is either a hundred times greater or a hundred times less than the two given estimates. Also, around 98 per cent of these are hydrogen (H) and helium (He) atoms in proportions slightly greater than 74 per cent for H and 25 per cent for He, with all other matter making up < 2 per cent. Helium atoms are effectively two protons plus two neutrons bound together.

Since there is a wide variation in the order between 10^{78} and 10^{82}, for the sake of simplicity and in view of that variation, let's consider that if all the matter in the universe consisted just of hydrogen (largely made up of one proton), then the mass of the universe in terms of that matter just using the hydrogen atom model may be calculated. The electron has virtually zero mass, and the dimension of a proton is given as 1×10^{-45} m^3. The total volume of mass in the universe would then be in the order of:

$$1 \times 10^{80} \times 1 \times 10^{-45} = 1 \times 10^{35} \text{ m}^3$$

Remember, this is a rough calculation based solely on the size of a proton, omitting the extra protons and neutrons making up the rest of matter, and may be inaccurate by a factor of plus or minus 10^2 as stated.

The figure 1×10^{80} mentioned above is the estimated total number of atoms in the entire observable universe and is still an extremely large figure, but it is small compared to the size of the universe if the size is based on expansion at light speed, in what I have described as the 'photonic universe', and in how many times this mass could be fitted within that total volume. Given that the natural shape of the universe is still undetermined as the universe is in isotropic expansion, it is likely to be an expanding sphere. If the universe is considered to be a sphere that has been expanding since the instant of the Big Bang at the speed of light, then using simple geometry, we can calculate the total volume in cubic metres as follows:

Time from Big Bang to now = 13.8 billion (13.8×10^9) years. There are 9.461×10^{15} metres in a light year.

This gives the universe a radius of:

$$13.8 \times 10^9 \times 9.461 \times 10^{15} = 1.306 \times 10^{26} \text{ m}$$

The volume of a sphere is found from the formula:

$V = 4/3\pi R^3 = 4/3\pi\,(1.306 \times 10^{26})^3 = 9.33 \times 10^{78}$ m^3 for the volume of the universe

The value 1×10^{35} m^3, approximately calculated as the physical mass (given earlier) as a proportion of the total volume, equals an ordinal of around 1.07×10^{-44}. This value is roughly in line with other estimates and shows that the universe is virtually empty space. The value of 10^{78} is the lower estimate of atoms in the universe and similar to the volumetric value. This would mean that universally there is an average density of, perhaps, around one thousand to a ten thousand atoms per cubic metre given the estimated range between 10^{78} and 10^{82}.

HOW BIG IS THE UNIVERSE?

The value of the speed of light is used to calculate the expanding size of the universe because electromagnetic energy emitted at the instant of the Big Bang would also be in the form of photons, both visible and invisible. They form part of our universe and should, in my opinion, always be included. And this agrees with other similar calculated dimensions in order to define the current radius. As these photons travel at light speed, it gives a maximum value to the current estimated width of the universe in its entirety at the instant of calculation. This is due to the universe's continuing expansion. However, a new estimate of more than 156 billion light years wide is now proposed by the astrophysicist Neil Cornish (2004). This contradicts the foregoing calculation, but it is reasoned that the expansion of the universe is likened to compound interest in isotropic expansion. Although photons have been travelling for just around 13.8 billion years, they are now 78 billion light years from their point of origin. Confusing! But a simple analogy given is that of compound interest, where each light year causes a little more expansion than the previous one. It is suggested by Cornish that if you think of just one year's bunch of photons radiating outwards when the universe was just one million light years wide, it would have expanded by one thousand light years when the universe was a thousand light years larger.

An anomaly that will be discussed in chapter 4 is that, in relative terms, nothing can exceed the speed of light, with the seeming exception of quantum entangled communication between photons, as discussed earlier in this chapter. This would mean that photons emitted from the Big Bang in one direction at the speed of light would be travelling at twice the speed of light with reference to those travelling in the opposite direction. If this is true for all points of the expanding sphere, then the universe would be expanding radially at twice light speed. Under current thinking this is an impossibility since nothing can travel or expand relatively at speeds greater than light speed. This is one of those times when speed, size, time, and space operate with different rules from that which we can observe from a relatively stationary position. These will also be discussed in Chapter 4 I have already speculated that if some of those initially emitted photons could have been split soon after the moment of the Big Bang at 180°, then it is possible that they still communicate with each other from opposite extremes of this photonic universe.

The preceding calculated size of the universe may seem laboured, and possibly highly inaccurate given the discussion in the previous paragraph, but it shows that the observable universe is virtually empty, and this emptiness may apply also to the proton. The volume of the proton may also be proportionately smaller if broken down to its incredibly small subatomic particles relative to the space these occupy. Then the volume of all the matter in the universe compressed down in to its subatomic particles without space in between them would probably fit into the space occupied by the volume of a single proton and hence result in a singularity of unfathomable density (see Figure 2.8).

Substance is defined in several dictionaries as 'the real physical matter of which a person or thing consists, and which has a tangible, solid presence'. Also, *matter* is defined as 'that which has mass'. In the instant of creation, energy must have been formed by a multitude of high-frequency and highly energetic vibrations. The frequency of these vibrations would have to be incredibly high at the start, with wavelengths still forming and lengthening

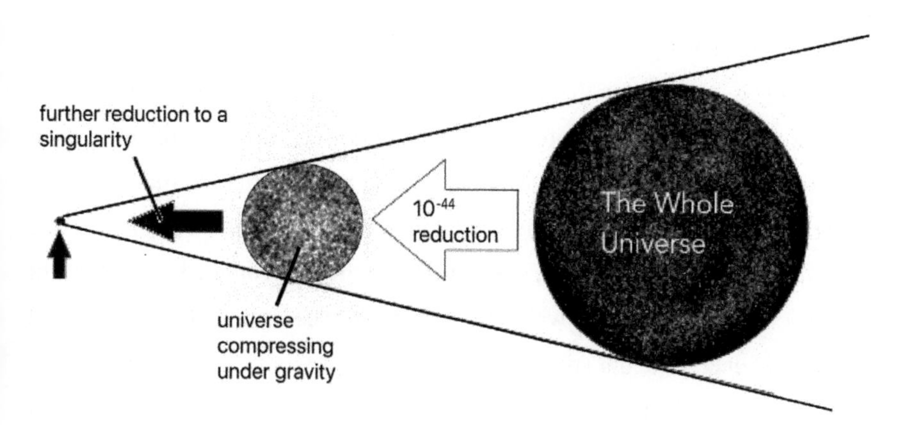

Figure 2.8 Reduction to a singularity. This illustrates the reduction from the whole volume of the universe to its matter and then all matter compressed down to a singularity.

over the next picoseconds as the energy spread in all directions.

In the maelstrom of that infinitesimally short time following the start, higher-frequency strands gave rise to the subatomic particles interacting with each other and with others of lower frequencies to form the basis of the elements that are recognisable as matter today. During the expansion, within the first trillion trillionth or less of a second of that initial Big Bang, the energy concentration could be likened to a tightly packed layer of plasmatic material resonating at extremely high frequencies, breaking down into small strands or strings that I would call 'stringlets' as they expand in all directions. My thoughts are that the stringlets would have a certain magnetic-like attraction to each other and with others of different, but harmonically related, frequencies, with those in close proximity forming entities to turn the frequencies of the strings into rotational spins. These would then begin to form the basis of subatomic particles that form all matter.

STRINGLETS

This book is written for the general reader, not those who are specialists in particle physics. My suggestion is that the oscillating frequency of stringlets is directly and dimensionally related to the time of release after the start of the Big Bang. My term *stringlets* is named in deference to those particle physicists who have developed the string theory of matter and the interconnectivity within the universe of all that matter. I would speculate that these stringlets are composed of a purely energetic, but still indefinable, flux that vibrates at frequencies that are difficult to comprehend and, since the frequency remains an unknown, is still a theoretical quantity and minute to the extreme, possibly being the smallest energetic formation of all. The energy from the Big Bang gives rise to the frequency at which these stringlets' oscillate, and these form the basis of possibly the highest frequencies in the universe. This energy may theoretically interact with, and be attracted to, other strings mainly of the same frequency in a sort of attractive resonance, and also with others of harmonics of that frequency, to form subatomic particles, many of which have been identified. This fits in with some interpretations of string theory. Before continuing, as I said earlier, I do not intend to go into deep nuclear physics, but I do intend to give a feel for the fact that there is a frequency component to all aspects of structures, from the incredibly small to the huge universe itself.

Now, if we accept that these stringlets interact with each other and as a result cause different frequencies, along with harmonics, that manifest themselves in forces such as gravity and nuclear forces, both strong and weak, and eventually cause magnetic fields as a by-product, then we have the building blocks for the components that make up atoms, which in turn make elements, compounds, atomic structures, and ultimately the visible universe and all that it contains, including you and me. Nuclear particle physicists theorise that a string would appear to be just like an ordinary particle (Wood 2019). It is thought to have a mass determined by its frequency and a specific charge, but the theorists also suggest it is a one-dimensional entity. But since frequency requires a time reference, this is a difficult concept. Viewed from our three-dimensional existence, this would appear to be an impossibility since a single dimension, x, would have zero length and zero thickness. In fact, even a two-dimensional entity, x, y, would still have zero thickness.

It is the combination of the second and third dimension, including x, y, and z, that gives tangible substance to matter formation. It is also a theoretical speculation that strings are impossible entities for us to contemplate, but at the moment of creation, where dimensions x, y, and z are yet to be formed, along with the fourth dimension of time, the concept of exactly what, where, and how energy existed at that instant is possibly beyond human comprehension at this time.

Strings, however formed, would have many other properties responsible for holding the fabric of the universe together, and one of the main properties is gravity, where strings attract all other strings in the universe in their own specific way. It may be the property of a certain frequency that gives rise to the theoretical graviton where a specific frequency, or combination with other frequencies, exhibits this universal force of attraction. Gravitons appear to be similar to other forms of energy and can be detected in waves. My own theories and comparisons between magnetism and gravity are discussed towards the end of this book in Chapter 11, where I state that the effects of this are felt across the universe. And where, I believe and theorise, those gravitons and photons have similar characteristics, but on different scales and with different frequency ranges and are not being closely linked.

GRAVITATIONAL WAVES

The gravitational waves discussed are not part of a basic frequency but a variation or modulation of this attractive force. Since gravity diminishes according to the inverse square law from a central individual mass, any mass in question would have to be very large in order to exert a gravitational pull over great distances, but collectively would add to other gravitational forces exerted by other massive sources. Variation in gravity waves may occur due to changes or physical vibrations in those masses. Research at the California Institute of Technology (2016) suggests that the frequency ranges start at kilometre wavelengths and proceed up to those approaching the size of the universe. This would further suggest that the entire universe shares a singularly common gravitational force in some way and that its variation gives rise to a frequency component that affects the whole of the universe. Variation in gravity caused at one side of the universe will ripple across to eventually be felt at the other. Gravity waves sensed on earth may have started out many millions of years ago, travelling at high speed.

The foregoing relates to a question which was posed by a group of three of us in 2010—Marjorie, my wife; our late friend Ray (to whom this book is dedicated), and me—at a regular discussion meeting, and subsequently placed on an internet astronomical enquiry site. The question was 'Would all the planets be instantly released from the gravitational hold of the sun and fly off tangentially if the sun disappeared in an instant?' This stimulated many discussions on the site with people from around the world and many groups getting involved with this question. Historically this question is not new with investigations into what was to become a vast area of interest starting out many years before.

Many groups reported back with increasingly complex discussions and formulae being presented as to whether gravity has speed. This result, highlighted by Siegel (2016), and from some international universities undertaking research, eventually led to the answer that planets would not be instantly released, not for around eight minutes as far as the earth is concerned, because the gravity is still an attractive force lingering for this time, exactly the same as the radiation from the sun. We on earth we would not see any effect for around eight minutes and twenty seconds, and more so for outer planets because gravitational effects also travel at the speed of light. However, gravity is not constant and can have frequency in the attractive force it exhibits. Gravitational waves or variations are now a subject of further research as previously mentioned.

THE BUILDING BLOCKS

At this point we now have the building blocks of substance from vibrations in the form of frequencies arising from the start of our particular universe's expansion cycle. String theories and the many frequency-related components, including muons, leptons, bosons, and neutrinos, led to quarks (of flavours up, down, charm, and strange, along with high and low nuclear forces). These make up hadrons, including two of the main and familiar components, protons and neutrons. These subatomic particles must again be defined by the energetic frequencies with 'flavours' that could be thought about as unique mixtures of string frequencies further defining the characteristics of the quarks and their interactions in the formation of fundamental particles. Although still in essence theoretical quantities, but mathematically proven, protons form the bulk of mass are coupled with electrons that are very much smaller from a negatively charged group called leptons. Forces produced by subatomic vibrations try to attract the two together and, in some cases, have succeeded in the moments following the Big Bang to specifically form the hadron fundamental particle, the neutron. The neutron is therefore a proton with a captured electron. It is probable that the energy contained within atoms is directly related to the spin speeds of the various components, and as we shall see, these rotational speeds are incredibly high.

Chapter 3

FUNDAMENTAL PARTICLES

In the previous chapter I discussed a theory of the formation of matter. This chapter looks at the nature of matter from a point of view of interactions between fundamental particles focussing on the electron, proton, and neutron and on how they exist in a totally frictionless environment.

ELECTRONS AND MAGNETIC FIELDS

Electrons are strange entities found in all matter. They are peculiar in that they appear to always be dynamic and can never be found as fixed or stationary. Another peculiarity is that they repel other electrons and so must exhibit a field around themselves that tries to maintain separation from other electrons. Electrons are found within all atoms and within nuclei, but mainly they are organised in 'clouds' that orbit atomic nuclei at different levels. The ones forming the outer levels can be made to leave the parent atom and to join in the outer level of an adjacent atom that had lost an electron.

If a large source of electrons, such as at the negative terminal of a battery, is connected to one end of a conducting wire with a connection to the positive side of the battery where there is a lack of electrons, then electrons will continuously 'leap' from atom to atom by flowing from negative to positive. This is the basis of an electric current. This pressure (voltage) may cause a closing of the natural separating distance between electrons, thus effectively bunching or compressing them. This causes a pressure reaction resulting in the field's extending at 90° to the flow because of the collective fields repelling each other and bulging outwards whilst the flow of electrons is maintained. A simple analogy of this sideways pressure to electron flow is of water flowing at pressure through an expandable rubber hose. The hose will expand sideways proportionate to the water pressure and effectively store some energy until the pressure is stopped. This energy is then put back into the water as an 'induced' pressure pulse. With magnetic fields being formed, the individual fields join up and will attempt to repel any other field not emanating from the same electron flow in the same direction. This is recognised as an electromagnetic field.

An idea attributed to Albert Einstein suggests that there is a time-related quantum explanation for the formation of magnetic fields. It follows that if electrons transit between atoms at speeds close to that of light, then the electrons will flow along a conductor in a different referential time frame from that in which we exist. This slowing down of time forms a time bulge from the electron grouping since in their time frame, more are present, flowing in the conductor, than in our time frame. Although further discussion on this theory is beyond the scope of this book, mentioning it serves to show the mysteries of these fundamental particles and their effects, which we take for granted in everyday life with our use of electricity.

Electrons are almost pure energy in that they have very little mass. The energy exhibited by an electron is in the form of a wave that can have two frequency components, rotational and orbital. The rotational or spin frequencies would have been established at the instant of the Big Bang. The orbital component is to what the electrons are attracted to, but not captured by the proton, so they orbit the proton at high rotational speed, balanced by distance and acceleration under this attraction. The frequency rate at which the proton is orbited by electrons, coupled with the electrons' own rotational frequency, effectively forms 'clouds' or 'shells' around the nucleus, causing an overall balancing of charge and attraction, rendering what is now a balanced atom. The energy of the electron and other similar subatomic particles such as the positron, a positively charged electron, is never diminished unless it collides with other atoms, where the collision energy is changed into photons. The positron is called an antielectron, being of the same mass and dimensions of an electron, but positively charged. It is a relatively stable entity released from nuclei, but when in collision with electrons, the effect of the very high spin speeds interacting annihilates both of them and produces a gamma ray photon as a result. The term *antimatter* may apply in part to the positron.

At this microscopic level there is absolutely zero friction of particles capable of absorbing the energy as would occur in the macro case of air friction slowing down physical movement in the atmosphere. The spin of electrons is a definitive characteristic and effectively eternal—that is for the rest of time—and are classified as 'up' if clockwise and 'down' if anticlockwise, but as will be discussed, there is a 'wobble', which is a simplified way of referring to a 'half integer' spin.

ELEMENTS AND PARTICLES

Going back to the Big Bang and the maelstrom of energy that followed, giving rise to protons, neutrons, and electrons, collisions between these components then further gave rise to the joining together of some of them. To recap, the most abundant atoms in the observable universe (around 98 per cent) are those of hydrogen and helium. These are composed of just one electron orbiting one proton for hydrogen, and two protons and two neutrons each orbited by two electrons for helium. Singlet hydrogen is the simplest one to form. Other elements formed when protons came together with other protons and neutrons in a variety of numbers, each grouping and then being held together by what is referred to as a 'strong nuclear force'. The distribution of elements appears to back up this theory, with these elements listed in any standard periodic table. The simpler elements formed more easily and therefore are more abundant than the heavier ones in a descending order of number, but ascending individually in size and weight. As mentioned, hydrogen is the simplest and therefore most abundant, consisting of 74 per cent of the material in the universe.

A single electron orbiting a single proton may cause an extremely high-frequency wobble equal to the frequency of the rate of rotation. This happens on a larger scale with the moon and the earth in that, although gravity attracts the moon, the moon also attracts the earth, causing a wobble as they both orbit the sun. Evidence of this wobble can be observed with the tides; the seas and oceans are effectively being pulled towards the moon. This is discussed further in Chapter 11.

In the case of hydrogen, this wobble is easily balanced out by sharing the electron of another hydrogen atom making up a more stable molecule. Because of the rate of rotation and frequency of electrons, along with the fact that hydrogen has only the ground-state innermost energy level, this sharing and balancing can only happen just once, thus limiting the orbit to two electrons in this first shell for each nucleus, caused by the orbital frequency of the electrons. Because electrons orbiting the nuclei in what is now a hydrogen molecule are negatively charged, there is a force of repulsion between the electrons. At this, the first shell energy level, there is room for just two electrons; therefore, pure stable hydrogen in molecular form is designated H_2. Of the pair of hydrogen nuclei forming a molecule, it is likely that because of the repulsion of the orbiting electrons away from each other, they maintain a balance, likely the result of their being always directly opposite to each other in the shared orbits, and a constant distance between them. This makes them slightly diamagnetic, but not sufficiently enough to form a solid matrix at temperatures slightly above absolute zero, but able to form liquid hydrogen at this very low temperature. However, hydrogen molecules can momentarily and uniformly align when subjected to an extremely powerful magnetic field. This property is exploited by MRI (magnetic resonance imaging) equipment.

ORBITAL FREQUENCIES

The rate of orbit around a nucleus is given by calculations from Niels Bohr's equations and Planck's constant as 6.56×10^{15} rotations per second. This is an incredibly high frequency, but not as high as the rate of spin of a nucleus at the theoretical circumference of its surface, or of the electron itself. This orbiting frequency, or spin, has been calculated as 2.98×10^{22} radians per second, giving a rotational speed of 3.64×10^{7} metres per second. This is about 12.1 per cent of the speed of light and gives a rotational frequency of 0.4768×10^{22} Hz.

Since scientific thinking is that the initial formation of hadrons occurred within the first instance of creation. These high frequencies would have had to develop within that time frame, hence the very high rotational rates mentioned above. The electrons themselves also have a spin in addition to the rotational orbital frequency. This is thought to be around 0.16×10^{32} Hz and to be in line with its almost zero mass, giving rise to the term *half-integer spin*. This term will be discussed later in relation to its application to neutrinos and other leptons.

If the electron is viewed as a small spherical entity spinning on its own axis, then the speed at its surface may exceed the speed of light. It is therefore possible that the volumetric size of an electron is related to, and limited by, the speed of light, and this could help determine the maximum dimensions possible for an electron thus determined within the parameters of the speed of light. An attempt to give a radius to an electron with the foregoing limitations could be calculated from simple geometric equations. The formula C (circumference) $= \pi d$, where C would equal one rotation and the circumference would be determined by the spin speed divided into the speed of light, $3 \times 10^{8} / 0.16 \times 10^{32}$, gives a circumference value of 1.875×10^{-22} m. And $2\pi \times$ the radius gives the circumference—therefore, circumference / 2π = the radius.

$$1.875 \times 10^{-22} / 2\pi = 5.968 \times 10^{-23} \text{ m.}$$

Imagining that the electron is a sphere, then the volume of the electron sphere would be $4/3\pi r^{3} = 0.283 \times 10^{-69}$ m^{3}. It should be noted that this volume may not fit in with other measurements that may be around, but it is just a suggestion that the speed of light is the main size-limiting factor.

In the early part of Chapter 2, a chronon was given as the time that it takes for light to transit the classical radius of an electron and was noted as being the smallest measurement of time. Using my simplified calculated value for the radius of an electron, based on the spin speed limited by the speed of light, this value divided by the speed of light gives another value for the chronon: $5.986 \times 10^{-23} / 3 \times 10^{8} = 1.9943 \times 10^{-31}$ seconds' duration—a shorter time period than the classical duration of 6.27×10^{-24} seconds. The classical value would give an electron a volume that is obviously much larger than my simply calculated value, but it would be spinning at a much lower rate if the surface speed is still related to the speed of light. The spin speed of the electron is amongst the highest known of all spin speeds, although there is nothing absolute at this time that has a frequency higher than this, but I speculate that the neutrino oscillation, to be discussed later, may be above this value, and another higher one may be manifested as the ultimate synchronised frequency in the universe, namely gravity.

Rather than being a particle, the electron is now more considered as a wave that curves around the nucleus where the frequency or rotational spin is, in fact, the frequency impressed on the rate of orbital rotation. To my mind this creates something of an enigma since there is a calculable mass of an electron given as:

$$9.10 \times 10^{-31} \text{ Kg}$$

so that it would appear to have the characteristics of being both a particle and a wave. An electron is said to have a half-integer spin, meaning that it takes two rotations (720°) to complete the cycle to get back to the starting point. It is theoretically likely that this causes another frequency characteristic at half the natural electron spin as it orbits the proton, causing it to have a precession forming the 'shell' or 'cloud' at specific energy levels. This is much like a photon but at a much higher wave frequency. Where electrons are in collision with atoms, causing the emission of photons, it may be that this sets the frequency of the emitted photon as a mixture, or harmonic, of the electron frequency and the orbital rotation frequency of the displaced electron. The generally accepted theory of photon emission will be discussed in the next chapter, which will link to the foregoing discussion.

This discussion on electron spin speeds is in the realm of investigation called 'hyper-physics', but electron spin ranks as the highest particle spin frequency so far encountered. These rotational rates are examples of the processes that created them from the incredible frequencies that were established at the moment of creation or within an extremely short time after the start of the Big Bang, as mentioned earlier. These rates also serve to show the extremely high amount of spinning energy as part of nuclear audit that is contained in single hadrons forming the nuclei making up all structures and the energy of the orbiting electrons. Positrons are also emitted from nuclei, particularly proton-rich radioactive beta, and have the same dimensions as electrons but with the opposite charge. They may also have the same peripheral surface spin speeds as electrons, but possibly in opposite directions, giving rise to being opposite in their charge characteristics.

BUILDING BLOCKS OF COMPOUNDS

Hydrogen atoms are amongst the building blocks of many compounds, and in singular atomic form they can borrow from, or share with, other, larger electron–proton formations to gain stability, not just with other hydrogen atoms. This is especially true where the outer shell of another atom is not full and would be capable of holding more electrons at a greater distance from the protons and neutrons forming the nucleus.

Because this sharing bonds two or more different elements together, although they are still molecules, the combinations formed are now called compounds because they are composed of more than one type of element. A hydrogen molecule is not a compound being composed of two of the same elements

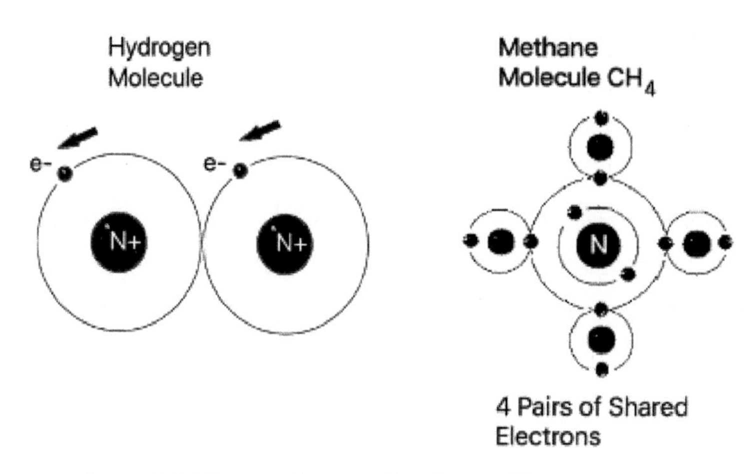

Figure 3.1 Electron sharing of carbon and hydrogen atoms.

(atoms) of hydrogen. The combination of carbon and hydrogen atoms also forms the basis for other familiar groups of compounds called hydrocarbons. A good example is where carbon, which has six electrons orbiting six protons and six neutrons, bonds with four hydrogen atoms (see Figure 3.1). This is designated CH_4 and is called methane. The outer shell of a carbon atom has just four electrons but could hold eight. By 'borrowing' four electrons from four hydrogen atoms, the carbon 'sees' that it has a stable, full shell of eight electrons, whereas each hydrogen atom 'sees' two electrons in its shell. For any individual atom, the maximum number of electrons allowed at each energy level or shell follows a simple formula of $2n^2$. This is where $n = 1$ and is the first (ground) level, $n = 2$ the second level, at a greater distance from the nucleus, $n = 3$, and so on. This gives a progression of 2, 8, 18, 32 … as shown in Table 1, which lists the maximum possible number of electrons for each cloud or energy level up to the sixth shell, but not all the shells are naturally full to these maximums. The ability of atoms to form molecules relies upon their ability to share electrons by trying to fill the outer shells in order to become more stable through electron or ionic bonding.

Energy level (n), also referred to as a 'shell'	$2n^2$	Total electrons possible in each energy level or shell
1 (ground state)	2×1^2	2
2	2×2^2	8
3	2×3^2	18
4	2×4^2	32
5	2×5^2	50
6	2×6^2	72

Table 1 Electron energy level maximums.

Visualising compound formations is difficult as the electron is so small, having an energetic mass of about 0.000 55 times that of a proton. It is shown greatly amplified in size for illustration purposes in Figure 3.1. Flat drawings can only ever be two-dimensional, so there is a need to visualise, with some imagination, that each circle is a sort of micro-subatomic ball where electrons orbit around the nucleus at incredible speed and precession to form a shell or cloud by being virtually everywhere around the nucleus at any instant. In sharing the electrons, they bond the compound tightly together and also allow different shell shapes depending upon the number of shells and subshells held by larger atoms.

ATOMIC WEIGHT

A standard comparison of the weight of atoms is called an atomic mass unit (AMU). This uses an unbound neutrally charged carbon atom as a reference. Pure carbon has six protons and six neutrons, with two electron shells containing six electrons—two electrons at the first energy level and four at the second—all orbiting the nucleus. This first level is called the 'ground state', and four electrons are at the next energy level. One twelfth the weight of this total mass is called an AMU and equals:

$$1.66053886 \times 10^{-12}\ \text{pg or}\ 1.66053886 \times 10^{-24}\ \text{g}$$

Since this amount is so small, the more usual reference is to use the abbreviation pg for picogram, which is 1×10^{-12} g.

An electron is approximately 0.000 55 of an AMU. The weight of an electron is approximately $9.132\,963\,73 \times 10^{-16}$ pg or $9.132\,963\,73 \times 10^{-28}$ g.

As the primary aim of this book is to look at the frequencies that make up existence, then further discussion on atomic structure, at this stage, would be too much of a diversion from this aim. The points discussed so far give a flavour as to how atoms bond with other atoms using sharing of their electrons to form those bonds throughout the complete range of known elements. The particles that make up these are atoms composed of eternally vibrating energetic entities (for the life of the universe). These gave rise to string theories. Putting a frequency value on all these vibrations is, at this time, an impossibility, but other frequencies emanating from changes in atoms can be calculated. The next chapter will start to develop my understanding of how interactions occur between atoms and start a discussion of some of the frequencies involved and their parts in the overall frequency spectrum.

Chapter 4

THE SPEED OF LIGHT AND ELECTROMAGNETIC RADIATION

The sun gives off many of the frequencies found in the electromagnetic spectrum, with each having different effects. They are all variations on a theme being that they all have frequency, as they are all electromagnetic. The other common attribute is that those frequencies emitted in the direction of the earth will arrive at the earth after just over eight minutes' transit time from the sun. This figure comes from a simple calculation of the mean distance from the sun to the earth divided by the speed of light, which gives:

149 597 870 / 299 792.458 = 499 seconds = 8.31 minutes or 8 minutes 19 seconds

Although this timing will vary slightly from apogee to perigee, the speed remains constant. The only variable with electromagnetic radiation is the frequency, starting from the very high (short wavelength) and decreasing in frequency to longer wavelengths, as shown in fig 4.1. These will be further discussed in this and following chapters.

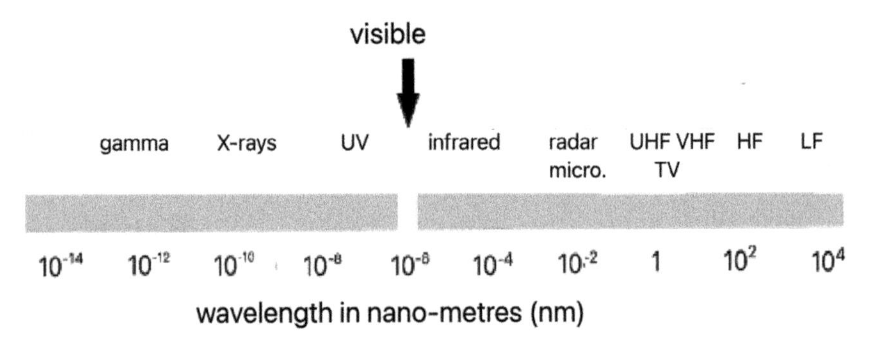

Figure 4.1 General electromagnetic spectrum.

All that we see and sense from the sun have two things in common, these being the speed of light and energetic entities previously alluded to called photons. Photons can be thought of as individual 'ripples in the ether' and are all very similar, differing only by their frequencies. Photons are, in general, generated and emitted from within atoms. If an atom is struck by a photon originating from elsewhere or hit by a colliding electron, a displacement of an electron orbiting within its shell may occur, due to the energy of the collision. The electron may be displaced to one of the higher-energy shells, resulting in an instability that will adjust itself, leading to the displaced electron returning to its original shell and emitting the energy gained on displacement as a photon.

FREQUENCY, AMPLITUDE AND ENERGY THEORY OF PHOTOS

The frequency, amplitude and amount of energy of the photon depends on its transit back to its original shell or energy level.

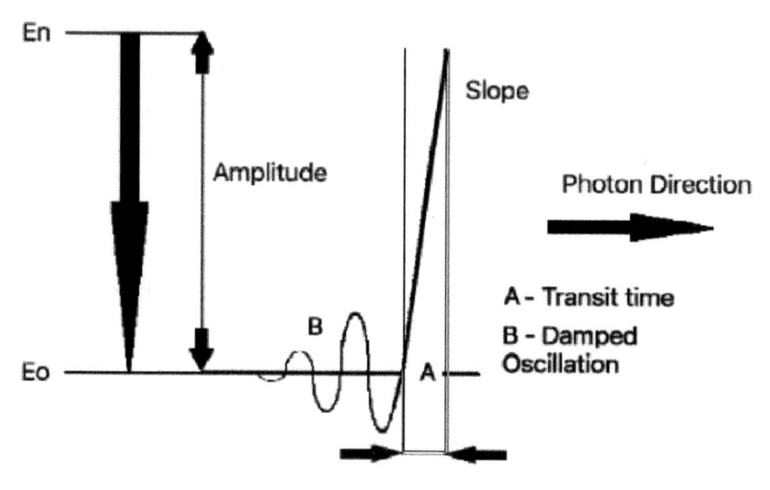

Figure 4.2 Photon emission theory.

Figure 4.2 is taken from my book *A Guide to Photo Therapy* (Somerville 2018) and is my theoretical interpretation of a photon emitted from an atom. E_0 and

E_n are the energy levels, where E_0 is the ground state and E_n the level that the electron is displaced to. The slope is the transit back to E_0, and A is the transit time. The pent-up energy gained in the initial displacement collision is released by the electron creating the photon as it transits. The amplitude, energy, and frequency are determined by the slope and the following damped oscillation, the oscillation being caused by inertia of the electron as it slightly overshoots the ground-state level. The amplitude of the photon would possibly be equal to the physical distance that the displaced electron travels back to its original state and could be a frequency-limiting factor for photons where a small displacement limits the energy released and therefore also limits the frequency. The duration would depend on the transit time and the total number of cycles forming the damped oscillation. It may also be that at lower energy displacements the electron takes a longer, more circuitous route, effectively spiralling back to its original level, in so doing lowering the emitted frequency and energy.

As has been discussed in a previous chapter, the spin speeds of electrons are extremely high, and it is likely that the energy tied up in their orbits causes an unbalancing of the position within a particular energy level if added to during a collision with another electron or a photon. It may be that the energy from this spin speed momentarily expels the electron from that level when a collision occurs as the electron takes energy from the colliding particle, forcing it to have a more widely dispersed and higher energy level. The reverse would then occur with the now unbalanced higher level ejecting the electron back to the balanced original level, emitting the excess energy as a photon.

Photons are differentiated by their frequencies; their energy is proportional to their frequency with higher-frequency photons carrying proportionately more energy than the lower-frequency ones. They all also have characteristics of being both a particle, because photons are given off individually, and as a waveform, that defines their frequency. They are not generally referred to by their amplitude, though this may depend upon whether the energy is released during the transit time of the electron when returning from the higher energy level, as shown in Figure 4.2, or whether a threshold has to be arrived at to release the energy as a common amplitude for all photons. If there is an amplitude to a photon, it may also depend on whether the returning electron spirals back to its original energy level or goes by a more direct path as discussed just above.

Photons may travel billions of miles across the universe for billions of years without losing their individual integrity and, in so doing, are able to provide information about their source. They are affected by gravity and beams of photons from far distant emitting stars are occasionally bent around massive bodies such as black holes, which are sources of high gravitational attractive force. These events are often seen by powerful telescopes and are called gravitational lenses. See Figure 4.3.

Figure 4.3 NASA image of a gravitational lens diverting the light from distant galaxies far beyond the gravitational anomaly. Some think it gives the impression of an alien face.

PHOTONS AND SPEED OF LIGHT

Photons are totally invisible to the human eye as vision is a construct of the brain and caused by an interaction with photons of a specific frequency band in the visible range by 'tuned' receptors. These receptors in the eyes give rise to the images we see, purely as an interpretation generated within the brain. The common link for all photons is that when they are generated and emitted from atomic structures, they travel away from the atom at the speed of light, whether across the universe or across the room. The speed of light is also a strange entity in that it is the ultimate speed in the universe, having a velocity of just under 300 000 000 metres per second or, more accurately, 299 792 kilometres per second or 186 282 miles per second.

Until the 1670s the speed of light was historically thought to be infinite. Then Ole Roemer, in 1676 (Soter and Tyson 2000), noticed that Jupiter's moon Io had apparent variations in its orbit thought to be the result of its elliptical orbit. He argued that the variation was not the result of the orbit but of the time that that light takes when Earth is in opposition to Jupiter, as opposed to when both are on the same side of the Sun. Light takes about seventeen minutes to traverse the diameter of earth's orbit path around the sun, so he argued that the time variation matched this value on account of light speed when in opposition.

Another significant aspect of the speed of light and photons is that if two photons pass each other from opposite directions, then the closing speed will be the speed of light and not be relatively twice it since the speed of light is the ultimate relative speed. This seems to defy logic in that if you drive a car down the road at 60 miles per hour (mph) and a car is approaching you directly in line from the opposite direction, also at the same speed, then logically on would expect the closing speed between the two of you to be 120 mph. However, the sound given off from the approaching cars will travel away from the vehicles at the speed of sound, and in this respect the sound has similar characteristics to emitted photons. The speed of the cars does not affect this.

Ripples in the Ether II

We can now consider sound energy transfer and some similarities with characteristics of photon transits. The atmosphere has a thixotropic rate linked directly to barometric pressure, and as with all liquids, this sets the speed of sound. This is a limiting factor, meaning that sound vibrations in air will leave the car at any instant at the speed of sound, not the speed of the car plus the speed of sound. This sound energy will also reach the ears of anyone, either stationary or in motion, at this set speed. Likewise, the sound from your car will reach the driver of the oncoming car at the speed of sound, timed from the instant of being emitted. The closing speed of the cars has no effect on the speed of the sound energy given off by either source as they approach each other, although the distance between sound wave origin and the receiver's ears may have shortened during the time that the sound wave was in transit. The only differences will be noted as changes in the frequencies of the sound waves, known as the Doppler effect. The same analogy works for two cars leaving in opposite directions and for a stationary listener, or one in motion, in any position in earshot of the cars at any angle from their trajectories. The speed of the sound arriving at the listener's ears will be constant regardless.

In the case of aircraft travelling at speeds greater than the speed of sound cause a shock wave is created that radiates at the speed of sound and is heard as a sonic boom. This is where that thixotropic rate of the atmosphere is exceeded, causing an effect that could be compared to cavitation which can occur in liquids that are accelerated beyond their own rate to a point where the water molecular cohesion breaks up. The general noise of the aircraft, along with the boom, will arrive at the speed of sound after the aircraft has passed. The relative speed of anything in motion to the speed of sound is referred to as the 'Mach number'.

So, how can we explain the closing speed enigma for photons? At an instant of time when atoms emit the photons, the speed at which the photons leave is at constant light speed. As in the car analogy, sound leaves at a constant speed of sound regardless of the physical motion of the cars, but unlike sound waves that can pass each other at twice their fixed transit speeds relative to each other, photons apparently do not. There is, however, a fundamental difference between sound energy and the energy of photons in that sound is radiated through the atmosphere largely in all directions, limited and dispersed by the very medium that radiates it.

Photons are emitted individually and in only one direction, in a straight line from the emitting source, and with few exceptions, maintain their individual integrity and require no limiting medium, as previously discussed. Photons are therefore enigmatic. My own approach to this relative light speed enigma is simplistic but fits in with Einstein's premise that relative time changes with speed. This is also known as 'time dilation', which is directly and proportionately related to the speed of light. This time dilation has substance, and has been proven by atomic clocks on orbiting satellites running microseconds slower than their static earthbound counterparts (Chou et al. 2010). By logically applying an extremely simple explanation based upon time dilation, we find that, for photons, their time frame slows down or dilates by half in relative terms to each other as the two travelling from opposite directions approach each other from proportionately different angles. This would then give the relative closing speed as the speed of light, thus maintaining the universal rule that nothing can travel relatively faster than the speed of light under any circumstances. If the two photons could pass each other at double the speed of light, then their relative mass would double in line with the frequency and energy carried, all with reference to each other. Since the energy/mass of a photon cannot be affected once the photon is emitted from a stationary source, it would be impossible for the photon to exceed the relative speed of light, so two photons passing at twice light speed could not relatively gain energy. The only answer to this enigma is that relative time slows down. The same would apply to photons travelling in opposite directions, being relatively unable to separate at speeds greater than that of light because that would entail a relative loss of energy and mass if separation were twice light speed.

I would also theorise those photons, or anything in motion, produce what I would term a 'chronular gravity field' around them. This is attracted to and 'quantumly' engages with other objects in relative motion. It would seem to me that gravity, distance, and time are closely related, with the only constant being the distance between two specific points of reference. The effect of these time fields would affect apparent time of those objects travelling towards each other. As individual chronular gravity fields approach, the relative fields become more intense and therefore proportionately slow apparent time down. This double time effect causes time dilation of the two fields where two objects approach directly or at relative angles. They appear to slow down relative time both in front of them and between them as they approach each other. As the objects travel away from each other's chronular gravity fields, the interactions would still be attractive, but each one extending time in a form of gravitational drag, weakening the chronular gravity field and thus increasing apparent time such that the objects depart from each other at velocities relative to the speed of light as a maximum and hence balance out energy and frequency changes. This time field theory would suggest that fields extend from every universal object and affect every other object in the same way that gravity does. If this is true, then it is possibly directly related to, and perhaps in some way part of, gravitational attraction as a whole.

In all cases photons will only be transmitted or received at the speed of light, thus changes to photons in transit must be caused by the relative motion and direction of the originating source. The previously mentioned Doppler effect increases the frequency of sound waves to a stationary listener ahead of the moving source and conversely reduces them when the source is moving away, but always maintains the speed of sound. This change in energy and characteristic equally applies to photons. The foregoing reasoning may provide a basis for a simple calculation showing by how much time slows down relative to the speed of any object or body in motion when viewed from a stationary point. If the object is travelling at around 60 miles per hour (100 kph), then time may slow down by this speed as a proportion of the speed of light. Simply applying this, and given that there are 3600 seconds in an hour, we get the speed of the object in metres per second = 100 000/3600 per second. Taking this value, relative time change would be:

$27.77 \text{ msec}^{1}/c = 27.77 / 300\,000\,000 = 0.000\,000\,092\,592$ times slower in time,

but the relative speed of light observed when moving remains constant.

This may be just a simple exercise that could also be subject to other factors, but as time slows for the photons, their speed observed from a stationary point of view remains the same. However, as discussed in the earlier, it also means that the frequency and energy of the photons is, in the same relative terms, also lower for those photons emitted backwards and travelling away from that stationary point. This is the basis of red shift measurements, discussed in Chapter 2. This change in observed frequency to lower than the original for the photons emitted from sources travelling away from the observer means that the photon has effectively changed in both its frequency and energy. The mass of the photon would also have changed in relative terms, becoming lower. The formula previously discussed gives the mass of a photon by multiplying Planck's constant by the frequency to give the energy, e, of a photon, with Einstein's $E = mc^2$, from which we have:

$$hv = mc^2, \text{ therefore } m \text{ (mass of a photon)} = hv/c^2.$$

This shows that mass (m) of the photon reduces in proportion to frequency change in relative terms as the photon recedes from a stationary observer; however, the speed of light remains constant regardless.

Ripples in the Ether II

The question then is what has happened to that loss of energy/mass that made up, say, a blue photon when, on account of the Doppler effect, it became a red photon or a photon of longer wavelength? I refer back to my additional way of describing kinetic energy: 'the dynamic relationship between anything in motion relative to anything stationary or in a lesser state of motion'. With this statement in mind, if a stationary observer were witness to a body emitting a photon as it travelled towards the stationary point, then a lower-frequency photon emitted from the moving object would experience a shift towards the blue end of the spectrum, thereby comparatively gaining energy and mass from the dynamic motion of the emitting source. This could mean that a red photon could become a blue or higher-frequency photon by gaining mass and that it would be observed as a higher frequency moving towards the blue end of the spectrum.

To sum up the foregoing discussion, it would suggest that for photons emitted in the opposite direction of travel from the source, energy is reduced in proportion to the departing velocity lost as a sort of energy-absorbing gravitational 'drag' from the emitting source, thus lowering the energy and frequency. Gravity is known to travel in its effect at light speed. Photons emitted in the direction of travel gain an energy boost, increasing in energy and frequency, from the source, again in proportion to the speed of the source. Since the speed of light cannot be changed, the only variable possible is the frequency. This in itself gives proof to the constancy of the speed of light.

To most people, including me, the nature of photons and the light speed enigma are perplexing. I certainly do not have all the answers. My opinions, from having studied physics as an undergraduate sufficiently enough to teach A-level students and other undergraduates, include an acceptance of the facts that time, distance, energy, and the speed of light are relative variables of the whole space-time continuum but, from our static observational position, seem fixed and constant. One can speculate that if you could travel at the speed of light along a photon beam, then the adjacent photons of light travelling in the same direction would logically be stationary to you. This would seem weird, as relativity theory suggests that as speed increases, time slows down or dilates, so therefore an anomaly exists such that time effectively stops. This would suggest that time would not exist for anything travelling at light speed relative to anything stationary.

For a living organism travelling at light speed and somehow being able to view earth as the stationary point from which it was travelling, everything on earth would appear to be frozen in time from the point the light speed was achieved. This may not be so strange or complex as it would simply mean that the observer, travelling at light speed away from a source, would effectively be viewing an instant in time like a photograph made up of photons or pixels all emitted at the same instant of time from the source and keeping up with the transit of those photons. The effect would be that the time view is frozen and never changes, much like a single frame of a movie. However, the hypothetical viewing of relatively stationary photons by anyone travelling at light speed with those photons may be impossible as the relative energy of the photons would be at a zero level with infinitely long wavelengths, meaning a frequency and mass of zero. They would be unable to stimulate any receptors in the eyes of the observer travelling at light speed. In fact, it would hold true that photons would not exist to anyone travelling along a beam of light as their wavelengths would be infinite and their energy zero.

This begs the final question on this subject of light speed and time: what would the observer travelling at light speed see on a stationary planet whilst travelling directly toward it? Basically, photons cannot travel towards our observer faster than the speed of light so visualising them would mean that time views may double. Things on the planet may seem to be moving at very high speed, like fast-forwarding a video. This would also be subject to the Doppler effect in that the frequencies of all photons coming from them would be shifted towards blue. The foregoing discussion would logically suggest that there is a difference in observed time dependent on whether travelling at the speed of light towards a stationary source of photons and if different from time observed when travelling away from a stationary source. However, this is purely referential, as in reality time would have neither slowed nor quickened, just being the observational illusion of the stationary point being viewed from a moving source.

Another aspect of the speed of light has been proposed, namely that any physical object that moves at any speed relative to a stationary point begins to convert its mass into energy. A simple analogy is to be found in travelling in a car: the car and the passengers gain potential energy, which changes to kinetic energy as the car slows down. Also, if we then consider the fact that the car is actually travelling over the curved surface of the earth, we conclude it may be that the car's relative mass becomes less as gravity is reduced as a fraction of the escape velocity. Escape velocity is about 18 000 miles per hour. Travelling at 60 miles per hour would possibly reduce the gravitational attraction by 60 / 18 000, which is 0.000 018 33. If travelling at the speed of light, all the mass of the object would possibly become pure energy. Lots of theories abound that mass becomes infinite at the speed of light, suggesting that the energy required to accelerate to such speed would eventually convert the energy to mass. In actuality no one knows, and all are just mathematical theories, but as all mass is interchangeable with energy and as energy is frequency, at the speed of light the mass converted to energy effectively becomes photons, one of only a few particles that truly travel at light speed, with photons being made up of energy strings of pure frequencies. These frequencies would be relative to the elements making up the original mass. The foregoing discussions are fully open to debate and simply reflect my own thoughts on the subject.

RADIOACTIVE ISOTOPES AND OTHER PARTICLES

At the start of Chapter 2, I described photons as 'ripples in the ether', the ether simply meaning the space through which they travel and not something that is known to have substance. These ripples are everywhere and are called electromagnetic radiation (EM). They are, as already discussed, caused by electron bombardment interacting with the structure of atoms or through the instability of unbalanced atoms, particularly in heavier elements. These elements are called radioactive isotopes and nucleotides, and, along with nuclear fission reactions, they give rise to the higher-frequency photons. There are other types of radiation composed of subatomic particles such as alpha radiation (protons and neutrons) and beta radiation (electrons).

Alpha radiation is composed of two protons and two neutrons that are bonded together and ejected from heavier, unstable elements such as uranium. The alpha particle is effectively the same as a helium nucleus and has a relatively slow speed of transit. Though damaging when directly impacting biological structures, they can be blocked by a sheet of paper. Beta particles are electrons ejected from a neutron which splits, leaving a proton and the ejected electron. There is no set speed at which they travel as this speed depends upon the release of energy of the decaying atom. This speed is termed 'Q value' and is very difficult to quantify.

There is, however, another elusive particle that is emitted as a result of a neutron's losing its tightly bound electron, the Fermionic neutrino. This may be from the binding energy arising from the strong nuclear force being released and would appear to travel at light speed. The Fermionic neutrino is not generally absorbed when encountering a solid but passes largely through unimpeded due to its size, but it is affected by gravity. The methods of detection are very complex and involve large detection tanks deep underground. It is estimated these neutrinos have a mass of less than one millionth of an electron. Neutrinos are stated to be the most abundant particle in the universe (Lesgourgues and

Verde 2019). It may be that they are a micro-version of a photon, having both particulate and frequency characteristics. As they are just one millionth the size of an electron; then, by ratio, the frequency is likely to be in the 10^{39}–10^{40} Hz range. They are further described as having characteristics of a half-integer spin in the same way as electrons as discussed earlier. In simple terms it means that it requires two angular rotations, that is 720°, for the neutrino to fully complete one cycle, returning to its original angular state at the start of each individual rotation. This would suggest to me that there is a wobble of the neutrino at a frequency of half the rate of the true spin. This may seem confusing, but the same half-integer spin theory is also applicable to other subatomic particles, such as leptons, and, when discussing frequencies, has to be included as part of the total frequency audit.

The binding energy released as radioactivity from other decaying atoms or particle annihilation of electron–proton collisions is in the form of electromagnetic radiation. It has a very high frequency and is designated as gamma (γ) radiation, having been discovered in 1900 by French chemist Paul Villard (L'Annunziata 2007).

GAMMA (Γ) RAYS

Gamma photons have the highest frequency of all known particles emitting electromagnetic radiation (EM) and the most highly energetic. This makes them dangerous to living tissue in that they are very penetrating. The energy from each photon impinging on tissue can disrupt or destroy molecular structures and can also damage the DNA structures within the cells formed by those molecules that make up all tissue. Gamma rays were initially thought to be a form of particle radiation, but it was found that they are without any specific charge. It became apparent that they were, in fact, electromagnetic in nature. The frequency of gamma radiation is between 3×10^{17} Hz and 1.24×10^{20} Hz (EHz to ZHz) (see frequency-related measurements in the Glossary).

In my other books I have often compared these large numbers used in describing frequency to the number of average-sized grains of sand on all the beaches of the earth. Scientists in Hawaii came up with the figure of around 7.5×10^{18} (Krulwich 2012) by observing and measuring all the dimensions of earth's beaches and averaging the size and volume of each grain of sand from samples, and from that calculating the approximate numbers on each beach, then adding this number to all other beaches to come up with the above figure. The rate of frequency oscillations contained within a photon of γ radiation, as so described in the upper range, would be many times greater than the number grains of sand if sustained for just one second. To further put this number into perspective, if one complete cycle of gamma radiation is drawn and is represented by a distance of just 1 cm from peak to peak, and if the cycles were to be repeatedly drawn continuously across a piece of paper, then just one second's worth would require the paper to be

$$1.24 \times 10^{20} \text{ (frequency in Hz)} / 1 \times 10^{5} \text{ (cm per km)} =$$
$$1.24 \text{ thousand trillion } (1.24 \times 10^{15}) \text{ km in length}$$

to include every cycle, or just over 0.77 thousand trillion miles, and would take many millions of years to complete. However, individual γ photons have a very short duration but are dangerous because of the sheer volume of photons emitted.

The energy contained in an individual photon is directly proportionate to the frequency or rate of change of electromagnetic flux. This is regardless of whether they are at the higher extremes of frequencies found in γ radiation or at much lower ones; this will be discussed in the following parts of this chapter, and continuing throughout the book. It is this rate of change that transfers energy when encountering objects or tissue whilst impacting at the speed of light. My basic reasoning for this is that electromagnetic photons are the product of electron displacements within atoms and have a frequency based on the energy of the displacement. When forcibly caused to interact with other electrons bonding and forming structures, they initiate a resonance that then transfers this high-frequency energy easily back into these molecular bonds. This causes them to be highly agitated, sufficiently enough to alter or destroy them. Photons are effectively the displacement energy from the emitting source, matching the energy levels in the atoms they encounter. Therefore, they may match in frequency and carry sufficient energy to be the cause of displacement of other electrons, thus causing photons to be released from within those structures. This is called the 'cascade effect'. This effect may cause different frequencies from the original initiating one to be emitted. This will be discussed later in the book. The very fact that gamma rays originate from the process of radioactive decay and have a higher frequency, carrying more energy to be able not just to displace electrons but also to destroy any molecular electron bonds they encounter, makes them highly penetrative and dangerous to tissue.

X-RAYS

This now leads to the next highest-frequency ray that are found, that is X-rays; the frequencies within the upper end of this range slightly overlapping and sharing the same qualities as the lower-frequency gamma rays but having different causes. These rays were termed *X-rays* by their discoverer, Wilhelm Roentgen, in 1895. At that time, they were unknown in any other context. They were described as 'X', or unknown, and were discovered as a side effect of Roentgen's experiments involving accelerating a stream of electrons in a vacuum chamber at a target. X-rays were found to be a by-product when electrons were accelerated from an element known as the cathode to a target called an anode with an electrical potential difference of around 25 000 volts (more modern X-ray devices use up to 1 000 000 volts). This potential difference was applied between these two electrodes, the positive being the anode and conversely the negative being at the cathode.

Figure 4.4 is a very simple illustration of an X-ray tube. The cathode assembly contains the heater and various electrodes that accelerate and focus the electron beam directed at the target anode. Accelerated by such a high voltage, the X-ray photons are caused to be emitted from the anode as the beam strikes it, as shown.

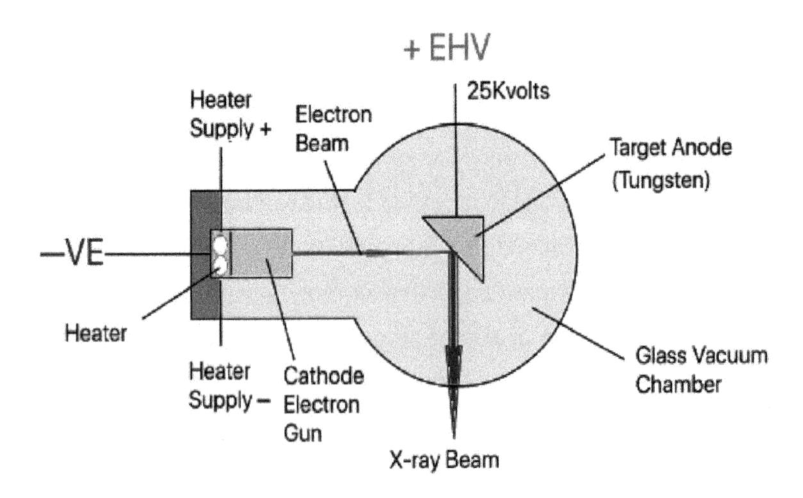

Figure 4.4 Simplified X-ray apparatus.

This high-acceleration energy causes the electron displacement between several energy levels in the targeted tungsten anode. When the displacement energy is released, it causes a very much higher frequency photon to be emitted as the electron returns to its original level. The discovery of, and the naming of, X-rays by Roentgen came about when he noticed that fluorescent-painted cards sited about one metre away from his equipment began to glow when these very high-voltage experiments were being carried out. Unsure of what was causing this glow, he coined the name *X-rays* for those hitherto unknown rays emitted from his experiment. Other scientists wanted him to name them 'Roentgen rays', but he declined. However, in some countries, they are still called Roentgen rays.

Roentgen later carried out experiments using X-rays beamed to pass through various substances to see how well these X-rays could penetrate the substances, which included his wife's hand, held over a photographic plate whilst being subjected to this radiation. The resulting image was of the bones in her hand, which were revealed when the plate was developed. On seeing the image, she remarked that she 'had seen [her] own death'. See original photo in Figure 4.5. X-rays easily penetrate through soft tissue but are absorbed by bone, which shows up as dark areas in developed photographic plates. Roentgen was awarded the Nobel Prize in 1901 for this discovery.

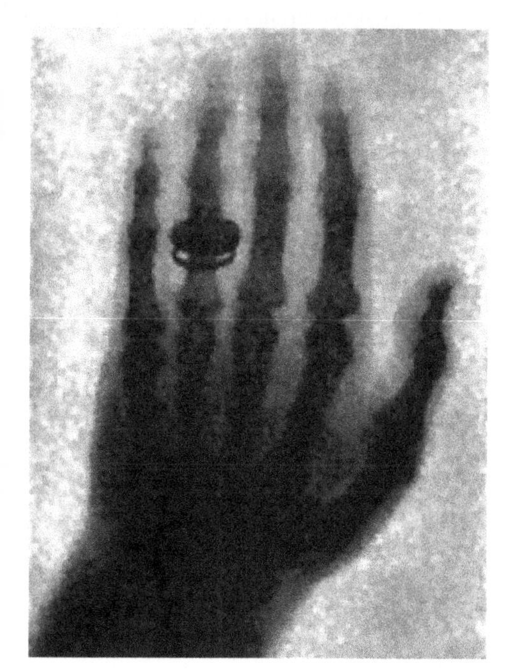

Figure 4.5 First ever X-ray image. (Public domain image.)

Early colour televisions that used cathode ray tubes were found to be emitting X-rays at initially dangerous levels. This was because the tube had three 'electron guns' carefully aligned to specific fluorescent phosphor dots on the screen that lit up when electrons struck them. Each electron gun beamed electrons through a shadow mask, allowing some to strike a specific dot aligned to the holes in the mask. These fluoresced and gave off red, blue, and green light depending upon the position of the electron guns. Four electromagnetic coils strategically placed on the outside of the tube and fed with 'sawtooth' waveforms caused varying magnetic fields that deflected the beams, causing them to scan and interlace, forming a raster. Beam intensity-controlling electrodes fed with the varying image information caused changes in intensities and synchronised to the raster formed a colour image that reproduced the one transmitted from the camera. The term *raster* means that a line is formed by the electron beams on the phosphor screen that steps down with each sweep, eventually returning to the stop to start again. Interlacing means that every second line is scanned in turn to present a smoother non-flicker image.

Because the three beams required for colour televisions needed to be accelerated to a higher degree than a single-beam black-and-white television using 15 000 volts, 25 000 volts was used, which is the electrical potential that will cause the emission of X-rays. This high-accelerating voltage of 25 000 volts is required for these colour televisions to work efficiently. At these high potentials, electrons were attracted to, and struck, the high-voltage-attracting anodes that then allowed some of electrons to form beams through small apertures. The absorbed ones gave off lower-energy X-rays, as did the ones that struck and were absorbed by the shadow mask. When the potential hazard to health was realised, television manufactures developed lead glass that absorbed X-rays, which was still used up to the end of the cathode ray tube television manufacturing process. Happily, modern LCD and LED digital colour televisions do not use such high voltages, so in this respect they are much safer.

X-rays are also emitted from extra-terrestrial sources such as the sun and also from cosmic rays originating from neutron stars thousands of light years away, but these are largely absorbed and scattered by the upper atmosphere. Observation of such X-ray sources requires satellite imaging equipment operating from satellites orbiting above the atmosphere. X-rays are the upper-frequency end of the photon family, where energy is emitted from electron displacements in atoms. Frequencies are in the 30 petahertz to 30 exahertz (3×10^{16} to 3×10^{19} Hz) range.

To recap, *all* photons in the X-ray frequency range and below are emitted as a result of electrons orbiting close to the nucleus and being unnaturally displaced by other high-energy electrons or photons colliding with them. The actual energy of an individual photon can be calculated. Max Planck (1858–1947) calculated a constant that can be applied to all photons to provide the energy value. If the frequency is known, then the formula $e =$ gives a value in joules per photon, where h is Planck's constant of $6.626\,070\,040(81) \times 10^{-34}$ and is the frequency. This formula has already been used previously. From this, we can start to give frequencies a comparative energy value, remembering that all energy came from the explosive start to the universe. The following are calculations using the formula to show energy in joules at its upper frequencies.

$$\text{Gamma photon} = 6.626\,070\,040(81) \times 10^{-34} \times 1.24 \times 10^{20}\ \text{Hz} = 34.727 \times 10^{-14}\ \text{J}$$

$$\text{X-ray photon} = 6.626\,070\,040(81) \times 10^{-34} \times 3.00 \times 10^{19}\ \text{Hz} = 19.878 \times 10^{-15}\ \text{J}$$

Perhaps a more user-friendly way of looking at energy, especially for the nonmathematical or non-scientific reader, is to convert to the more familiar watts value. This is very simple in that 1 joule per second of energy transferred is equal to 1 watt. This gives an idea of the quantity of photons required to transfer just 1 joule of energy. From the foregoing calculations for the upper-frequency X-ray photons, then, $1 / 19.878 \times 10^{-15}$ J = 50 306 871 918 704 photons emitted per second to transfer 1 watt per second.

The gamma and X-ray frequency energies shown are calculated from the upper level of the range of frequencies for gamma and X-rays respectively as previously discussed. They would have an overlap between the lower end and upper end of the ranges respectively, making them effectively the same at the crossover point. From these calculations, therefore, it can be seen that energy is directly proportional to frequencies and reduced by a factor of more than 10 from gamma rays to X-rays at the same relative part of each spectrum. The foregoing values may seem incredibly low, but it is the additive effect of the sheer volume of photons emitted and the length of time of exposure that can be dangerous to biological life forms.

Gamma rays and X-rays, electromagnetic in nature, are just the starting points on our journey down the frequency range. A standard way of describing frequencies within the electromagnetic band, including the frequencies already discussed, is to state the length of the wave. Wavelength (symbol λ) is given in relation to how many cycles pass a stationary point whilst travelling at the speed of light for a given frequency ($f = c / \lambda$). This speed is standard for all radiated and transmitted electromagnetic energy. When we look at the much lower frequencies, we see that the wavelengths become extremely large and eventually there comes a point when radiation as such ceases and photons are no longer the main carrier of electromagnetic energy. At this point we will still carry on investigating the radiated energy from very much lower frequencies with longer wavelengths, but it is not my intention to limit the book to electromagnetism. I intend to go beyond the electromagnetic into the audio-mechanical range, passing through to what I would describe as the 'sub-unity' range, meaning less than one hertz as previously described.

The shortest of all wavelengths are gamma waves, measuring from around 1×10^{-6} nm (1×10^{-15} m). This method of defining frequency by the length of a single wave is standard for electromagnetic radiation because the speed of light is constant for all photons and transmissions in a vacuum. However, the figure used to calculate the wavelength is rounded up to 300 000 000 metres per second, the actual speed of light being 299 792 000 metres per second. From here on, the wavelength will also be used as the standard reference, along with associated frequencies whose wavelengths get longer as the frequencies reduce within the electromagnetic spectrum.

Although both gamma rays and X-rays are dangerous if overexposure occurs to healthy biological tissue, they also have their uses in medicine and in this context can be life-saving. X-rays have obviously developed in their ability to image tissue densities. Both gamma rays and X-rays can be focussed such that they can be concentrated and targeted on specific points such as tumour sites. The concentrated energy available at the point of focus can be sufficient to destroy deep-set tumours. One of the first devices designed for such purposes was called the 'Cobalt Bomb' or 'Cancer Bomb' and was developed by medical physicist Dr Harold Johns and his graduate students in 1951 at the University of Saskatchewan. They utilised cobalt-60 to radiate and focus homogenous gamma rays in order to specifically destroy cancers. Cobalt-60 is an artificial source of gamma rays and is produced by exposure to beta radiation from atomic reactor piles.

The term *ionising radiation* is given to the higher-frequency gamma and X-ray photons because, being of high frequency and therefore highly energetic, they are able pass enough of their energy into any atoms they encounter, to cause not only the displacement of electrons within atoms they collided with, but also the complete loss of an electron from one of the electron shells orbiting the nucleus. This loss unbalances the neutral state of the atom affected, causing it to become a positive ion, hence the term *ionisation*. If the atom is part of a specific molecular structure, then ionising radiation will cause the whole molecule to become ionised, that is positively charged. If enough higher-energy photons impact on biological molecular structures, then the damage sustained by ionisation may be permanent. This can be carcinogenic if ionised damage is caused to DNA bases. Although the term *ionising radiation* applies to gamma rays and X-rays, the upper frequencies of the next in the series, ultraviolet rays, may also be ionising, although as we progress further down the frequencies, the problem of ionisation diminishes as the wavelengths become longer. This will be covered in the next chapter.

Chapter 5

LET THERE BE LIGHT! FROM ULTRAVIOLET TO VISIBLE RED

Light is generally referred to as encompassing the frequencies within the visible spectrum, starting at around 380 nm and going up to 700 nm wavelengths. Different wavelengths within each part of the visible spectrum have different effects on life forms and give depth and meaning to the physical world all around, along with an almost infinite variety of colours. All light consists of photons, and all photons originate from within atomic structures as previously discussed.

The main source of photons that light up the earth is the sun, but other sources of photons originate both from the earth itself and from outer space. The maelstrom of energetic activity throughout the body of the sun gives rise to its surface temperature of around 5500 °C to 6000 °C. This causes the radiation of particles and virtually all photon frequencies of the electromagnetic spectrum. The visible part of the electromagnetic spectrum is shown in Figure 5.1.

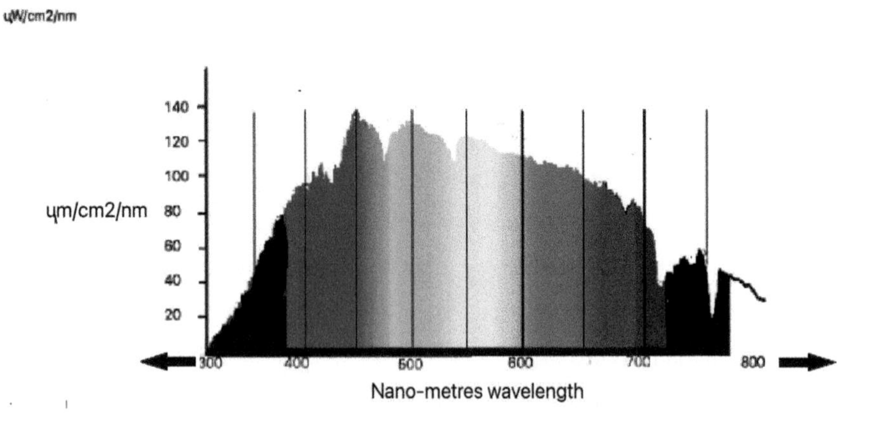

Figure 5.1 Spectral emission from the sun.

As may have been deduced so far, there is an overlap between gamma rays and X-rays. The ultraviolet radiation at 10 nm to 100 nm in wavelength is slightly longer than the lowest-frequency X-ray range.

UVA, UVB AND UVC

Ultraviolet (UV) is still a very high-frequency source of photons and can cause similar damage and problems as X-rays, but UV rays are less penetrating of tissue given their lower energy. They can be produced by both natural and artificial sources, the sun being one of the main natural sources. Fortunately for all mammalian life forms exposed directly to the sun, UV, like both gamma rays and X-rays, is largely absorbed and scattered in the upper atmosphere. However, some of the sun's UV rays of all wavelengths do manage to get through, along with other electromagnetic (EM) radiation. Ultraviolet radiation covers a wide waveband and is largely divided into three categories: UVA, UVB, and UVC. All UV wavelengths have some ability to ionise atoms within biological molecular structures, though this occurs more quickly and dangerously at the shorter wavelength. UVC at the upper-frequency (shorter wavelength) end of the UV band has a more effective ionising ability, making it more dangerous to tissue within a short exposure period. UVC and UVB can still be produced artificially from high-voltage discharge devices such as mercury discharge tubes and can also arise as a side effect of some high-temperature welding techniques.

UVC does have its uses. These were recognised in the late 1900s by Niels Finson (Gotzsche 2011) as being antibacterial. Finson used UV extensively in his research for which he was awarded the Nobel Prize. The energy that is carried in high-frequency ultraviolet photons is sufficient to disrupt and break down the molecular bonds, in particular of bacterial structures whose DNA is not nucleated but fills the main body of the bacteria. Penetration of tissue by ultraviolet frequencies is very superficial in that most is scattered at the skin surface, where it can cause damage to skin cells as well as bacterial cells if overexposed. This mainly occurs with UVC and UVB. The Australian Cancer Council website states: 'Exposure to UV radiation is the main factor that causes skin cells to become cancer cells' (Cancer Council New South Wales 2020). Australia has been described as the world's skin cancer capital. Ninety-nine per cent of nonmelanoma skin cancers and around 95 per cent of melanomas arise from excessive UV radiation from exposure to the sun and to artificial sources such as sunbeds and solariums.

ARTIFICIALLY GENERATED SOURCES OF UV RADIATION

This includes gas discharge tubes, where ultraviolet photons are a product of passing an electric current through and along the tube that contains specific gases. Fluorescent lights work by having a coating on the inside of the tube which gives off light as the current passes through and along it, and this generates UV photons from electron collisions with the gaseous mercury vapour molecules. When emitted, the UV photons strike the phosphor coating. It then fluoresces, giving off a bright white light along its length. The displacement of electrons in the atoms of the phosphor coating releases photons of different wavelengths within the visible spectrum; hence, since the eye interprets the variety of colours as seen simultaneously, they appear as white. When subjected to UV radiation, phosphor in fluorescent tubes absorbs most of the ultraviolet photons; therefore, most, but not all, of the UV radiation is contained within the tube.

Some sunbeds utilise standard fluorescent tubes because sufficient UV is emitted to cause tanning. It has to be stated at the time of writing that many lighting fluorescent tubes are being replaced by light-emitting diode strips resembling the older tubes. These do not generally emit ultraviolet radiation and require much lower currents to operate. UV radiation from the original types of fluorescent tubes, when mounted at some height above offices and other workplaces for the simple purpose of illumination, is at safe levels after a very short distance.

Lower-frequency UV sources are used as 'black lights', which use ultraviolet tubes without phosphor coatings. The term *black light* seems like an oxymoron, but it simply means that the light emitted is not in the visible range but that the effect it has causes fluorescence, where certain paints and objects, including bacteria, begin to glow when bathed in this 'light'. Humans have the ability to see certain ultraviolet frequencies when the lens of the eye has been removed. The lens filters out this higher frequency under normal circumstances. Using special tubes or devices that allow more UV photons to be radiated freely will cause fluorescence to occur when certain objects or chemicals are exposed to this radiation. Certain types of enzymes fluoresce when included in some washing powders. After washing, this allows the clothes to take some of the enzymes on board, making them appear brighter when subjected to the UV in sunlight, and causing them to glow when the wearer is in a darkened room with active black-light emitters. This is typical of dance halls and clubs where the use of such devices can be found.

ISSUES OF EXPOSURE

It should be emphasized at this point that UVA, although carrying less energy than UVB and UVC, is still carcinogenic if overexposure occurs either from sunbeds or sunbathing for extended periods without protective creams being applied, as mentioned above in the Australian article.

This begs the question as to why humans, who have lived for thousands of years in areas or regions of the earth where natural UV is greater than in other areas, namely in the tropics, where the area per square meter of the sun's radiation is more intense, do not seem to suffer the same exposure problems as people who do not originate from there. Melanocytes reside in the skin and in the eyes. They produce coloured pigments of melanin. Production in skin is increased by exposure to UV radiation and is a natural sunscreen against UV. Those who have their ancestry in these areas of higher UV radiation have genetically developed larger amounts of melanin, making their skin of a darker colour, and are largely brown-eyed. Whilst the production of melanin is increased in people with fair skin and causes short-term tanning, it is not sufficient to be a safe sunscreen against UV, so additional sunscreen application in the form of creams is required. Failure to be safely protected leads to ionisation of basal and squamous cells, and hence these can become cancerous.

Earlier, in the Introduction, I stated that we are products of our environment. Ultraviolet radiation is a very visible testament to that statement. Furthermore, the downside for dark skin coloured people living in areas of lowered natural ultraviolet radiation means that their skin pigmentation blocks UV that is required (see below) by the body lowering it to a level such that it affects the synthesis of calcium and may lead to problems such as sickle cell anaemia and reducing the immune system's ability to overcome disease.

A positive side of UVB radiation is that limited exposure to natural UVB interacts with types of cholecalciferol to produce vitamin D. This vitamin is responsible for regulating calcium in the body, which is referred to as calcium homeostasis. Calcium is vital for many functions of the body, including maintenance of muscles and bone, as discussed below. Vitamin D is found naturally in some foods and can be supplemented in several ways. Vitamin D comes in two forms:

1. Vitamin D_3 (cholecalciferol), which is produced (synthesized) in human skin by exposure to sunlight from a form of cholesterol called 7-dehydrocholesterol. Specifically, it utilises the small amount of UVB radiation that still manages to make it through the upper atmosphere after being largely absorbed by the atmosphere.

2. Vitamin D_2 (ergocalciferol). This is a form of vitamin D arising out of photosynthesis in plants, edible fungi, and yeasts. In the form of yeast, it is often used to supplement foodstuffs where natural exposure to sunlight is limited.

The lack of exposure to natural sunlight can be a source of problems such as rickets and osteoporosis, especially for those whose cultures require the total covering up of the skin at all times when outside. Vitamin D_3 is one of the essential soluble fats known as secosteroids. These play a part in the efficient absorption of calcium in the intestines, along with magnesium and phosphates. They are involved in the process of bone maintenance and muscle contractions. In children, a lack of vitamin D leads to a softening of bone, which can result in pathological fractures. In adults the lack of transportable calcium causes depletion of cancellous bone, particularly in the spine and around major joints. Most of the D_2 and D_3 in the human body is synthesized by sunlight.

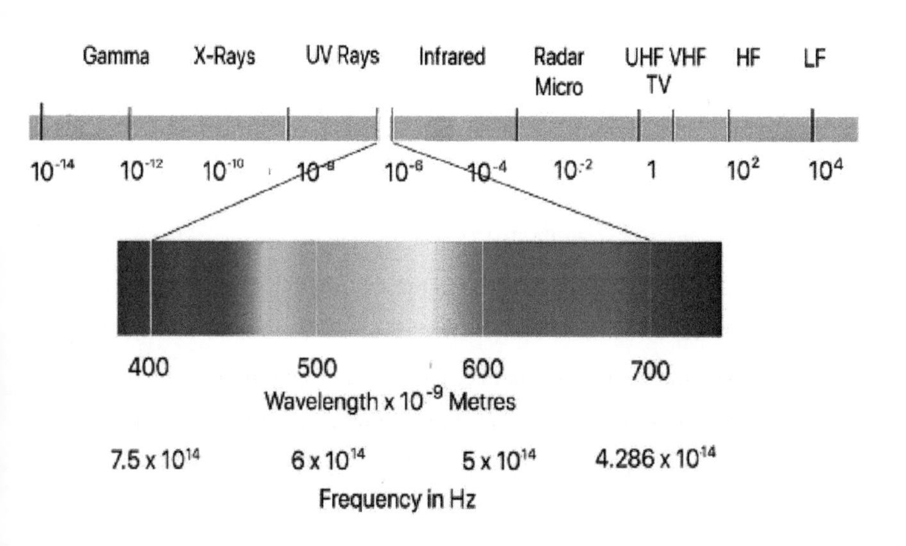

Figure 5.2 Visible light and its place in the electromagnetic spectrum.

At this point descending through the frequencies, we enter the visible part of the electromagnetic spectrum. This small group of wavelengths represents less than a trillionth of the electromagnetic spectrum as a whole and is where the effect of photons begins to become visible to the human eye. See Figure 5.2. The longest wavelengths of UV start to appear a deep violet to the observer. These merge to eventually become a deep blue at around 400 nm.

Starting at around 400 nm in wavelength, blue has had a significant signature in the development of life. Earlier in this chapter, the use of UV to kill bacteria was discussed. The UV used for this purpose is from man-made sources since most naturally occurring UV sufficient to be highly antibacterial is absorbed by the atmosphere, as was also discussed.

PHOTON RESONANCE AND BACTERIA

My reason for saying that ultraviolet and blue light appears to be antibacterial may be supported by looking at prehistory. Bacteria were developing in the Precambrian period of 400 million to 500 million years ago. They were still restricted from advanced development or evolution by the sun's radiation, which allowed certain high-frequency light wavelengths to penetrate through the atmosphere down to the earth's surface, without being absorbed or scattered by that early atmosphere. I suggested some time back, and have written about this in other books (Somerville 2017), that the energy of blue photons, though not as high as that of UV, has wavelengths that match dimensionally the size of some bacterial structures, either directly or as a multiple of wavelengths. This introduces the idea of resonance, where electromagnetic radiation will be absorbed and built up, or resonate, as the photons are absorbed, with some being reflected back and forth internally within the bacterial structures they irradiate. This is like blowing over the top of a bottle, causing it to sound a whistle or produce a tone, dependent upon the internal dimensions, that begins to resonate as the internal pressure pulses back and forth. This analogy will have relevance in the next chapter when looking at the generation of microwaves.

Back in the Precambrian period, bacterial photon resonance could have caused the build-up of electromagnetic energy in and around the DNA structures contained within the prokaryotic bacterium, eventually reaching a threshold of concentrated energy where damage to the DNA's internal structures may have occurred. This may have been especially so around the non-nuclear DNA strands that are found throughout the body of the bacteria. UV radiation, having a higher frequency and therefore proportionately higher energy, would possibly kill bacteria directly if exposed without atmospheric absorption of the photons and without the need for resonance build-up. As the atmosphere developed, more UV absorption began to take place by the atmosphere, but the lower-frequency photons in the blue part of the spectrum still got through.

At the time of the Precambrian period, the atmosphere was stabilising, and particulate matter with physical dimensions around one micrometre also stabilised and spread throughout the upper atmosphere. These were of a size that specifically begins to scatter blue-frequency photons arriving as part of the spectrum of solar radiation by diffraction. This process is called 'Rayleigh scattering' and is responsible for the blue sky we see today. See Figure 5.3.

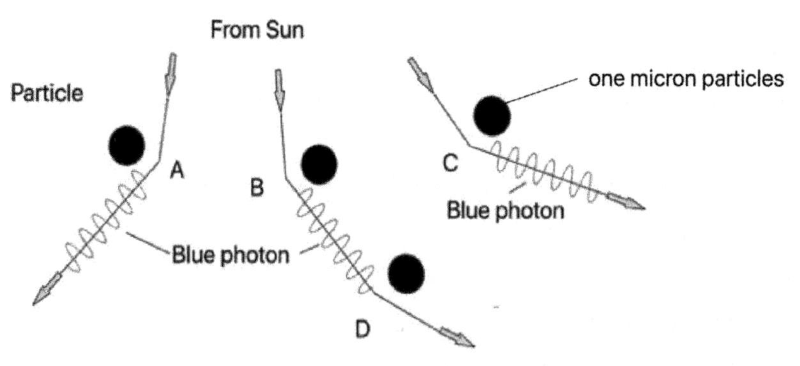

Figure 5.3 Rayleigh scattering.

This shows that there are multiple angles that the photons go through when encountering multiple particles. Whilst the diagram shows a generally diffused downward trend, there are many photons that are scattered back up into space, which gives rise to Earth's being called 'the blue planet' when viewed from space. This changed in the Precambrian atmosphere, causing scattering of most blue-light photons, which then allowed bacteria to develop relatively unimpeded. These bacteria built up stromatolites, formations made up of cyanobacteria. Viewing the spectrum of the sun's radiation, we see the peak value falls in the middle of the blue part of the spectrum at around 460–470 nm. It is significant that this frequency is now used in devices that help stop the proliferation of bacteria in and around healing lesions.

NEONATAL JAUNDICE

Another medical use of blue light is in the treatment of neonatal jaundice. Premature and some new-born infants develop jaundice as a result of their livers not being sufficiently developed to be able to process serum bilirubin. This makes the skin turn yellow. Exposing the infants to blue-frequency light in specially designed incubators causes isomerisation of bilirubin, altering the molecular structure in the superficial veins to a form that can be processed by the liver. Previously, fluorescent strip lights were used that had the same effect on account of the blue and UV components being present. This effect was discovered after the accidental exposure of test samples of affected infant blood to direct sunlight. When the samples were tested, this change in bilirubin was noted (Cremer et al. 1958).

PHOTONS AND VISION

As we have descended the frequency range into the visible spectrum, we see how resonance begins to help identify the different wavelengths. All photons are invisible to the eye, and thus all colours are a construct of our brains.

The human eye has specific colour receptors categorised as short, medium, and long. These are called cones because of their structural shape and are individually wide-spectrum 'tuned' to be stimulated by three overlapping colour bands of blue, green, and red light. These are the short-, medium-, and long-wavelength parts of the visible spectrum. Action potentials generated by the cones of the eye are, in essence, all the same but are identified by origin and interpreted by the brain in the visual cortex as colours. Although positional and motional pre-processing occurs at the retina, ratios and mixtures of the three primary colours signals from the red, blue, and green wavelengths give rise to the infinite variation across the visible spectrum. There are approximately six million cones in each human eye. See Figure 5.4.

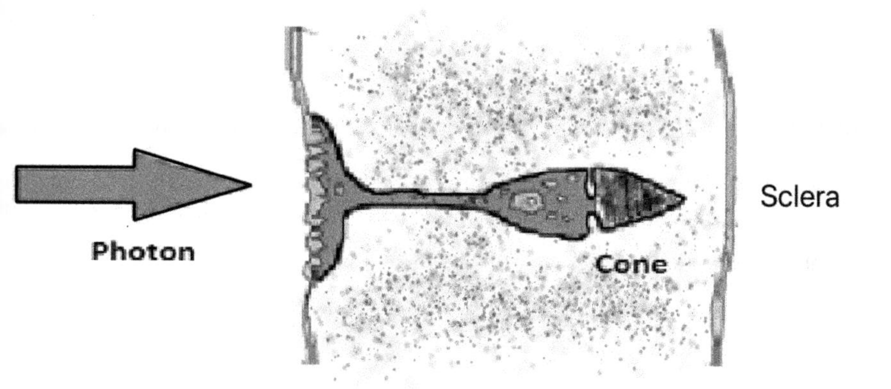

Figure 5.4 Typical cone cell colour receptor.

These cones may also possibly be categorised as small, medium, and large in relative size and as such may build up energy because they respond, or begin to resonate, at corresponding short, medium, and long wavelengths. Stimulation has a peak response at the corresponding frequencies of the photons whose wavelengths are of similar dimensions to that of the cones. See Figure 5.5.

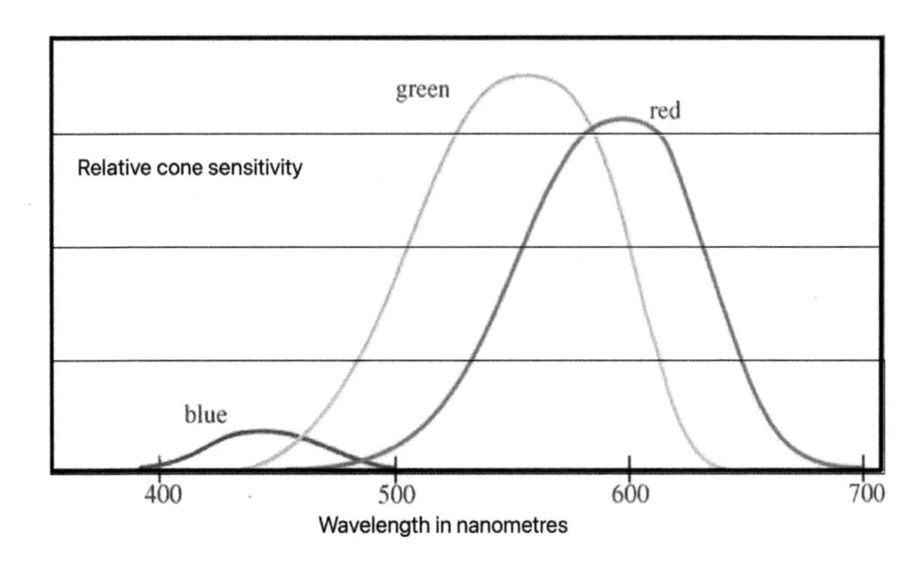

Figure 5.5 Spectral sensitivity of human eye receptors.

Photopsins are light-sensitive proteins found within each receptor cell that are affected by photons entering them. Rhodopsin, also known as visual purple, is a biological pigment. When exposed to light, it effectively changes the characteristics of the cone cell by isomerisation and gives rise to a chain of events that lead to the generation of an action potential. Rhodopsin is not naturally produced in the body and has to be synthesized from beta-carotene, a pigment typically found in carrots, giving some credence to the idea that carrot eating is essential for good eyesight.

The amount of stimulation of the cones causes them to send millions of signals along the optic nerves to the hindbrain occipital lobe, where the different rates and sources of stimulation are interpreted as the variety of colours. Rod receptors, called rods because of their shape, are far more numerous, amounting to around one hundred and twenty million per eye. These mainly function in situations with lower light intensity. Of course, even at night, there are many visible photons around, but the number is greatly reduced in comparison to in the daylight. Rod cells have a peak sensitivity of 507 nm and are many times more sensitive than cone cells. However, the rhodopsin, coloured purple, within them is effectively destroyed in bright light by bleaching and becoming clear, meaning that the rods only function in darker environments and provide images to the brain without colour. The eye, containing a greater number of rod receptors than cones, is stimulated by all the visible photons. As the amount of light gets less and less, only the rods function, meaning that at a specific intensity threshold the cones are no longer stimulated and vision then becomes colourless—black, white, and greys—as mentioned above. Photons of light that are not absorbed by either the rods or the cones are absorbed into the black coating of the sclera, also called the 'tunica negra'.

SENSITIVITY OF THE EYE

Since photons of the colour blue are of a higher frequency than red photons, the energy is proportionately higher. The sensitivity of the eyes varies to take this into account. It is also true that high numbers of photons concentrated on a single point can be damaging no matter what the frequency. This occurrence is possibly from overexposure, either directly from the sun or from artificial sources such as light-emitting diodes and laser devices. For this reason, artificially generated light intensity is classified as class I, IIA, IIB, IIIA, IIIB, and IV. These are rated in terms of the eye damage that can be done within the blink of an eye and were introduced because of the development and availability of lasers, particularly in the form of laser pointers.

Ratings follow a series using Roman numerals. Generalising the effects of laser or bright light sources is as follows: with class I to class IIB, no permanent damage will occur if momentarily exposed. Class III and above will cause permanent damage if any exposure directly into the eyes occurs, even if momentarily exposed within a blink response. Class IV is capable of causing instantaneous damage resulting in blindness and also ignition to flammable objects. This high-powered class extends to more powerful lasers capable of melting metal and commonly used in industry. It should be noted that even at the class I or class II level, sustained direct exposure to the eyes and tissue will still cause damage and should be avoided.

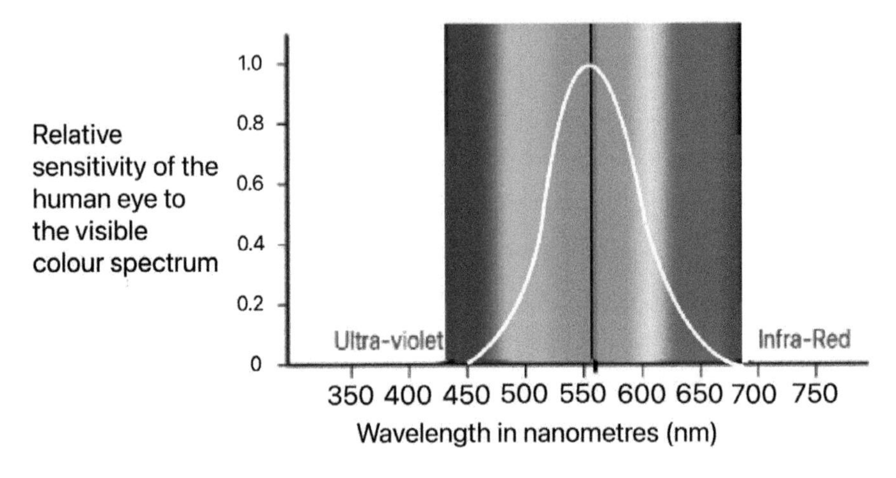

Figure 5.6 Overall spectral sensitivity of the human eye.

Figure 5.6 represents the mean sensitivity of human eyes in relative terms. The very high-energy blue end of the spectrum has the least sensitivity, whereas the highest is within the green band. From this point, as the sensitivity gradually lessens with increasing wavelengths, it reduces to zero as far as optical sensitivity is concerned at the start of the infrared 700 nm. A possible reason that sensitivity to blue wavelengths is much reduced is that blue wavelengths are still dangerous to the eye if the eyes are overexposed. Continued exposure causes retinal damage by way of the collective higher energy potentially damaging the cone receptors.

PHOTOSYNTHESIS

Continuing through the spectrum, the blue merges into green wavelengths. In the case of plants, the green frequencies are not particularly absorbed but are scattered. Photosynthesis absorbs photonic energies from the sun or artificial sources of nearly all frequencies other than green. Specifically, because chlorophyll is pigmented green, it absorbs mainly red photons, leaving the green to be reflected and scattered or passed through. The energy in the lower-frequency red wavelengths at intensities radiated down to the earth from the sun is at just the correct level for the photosynthesis process to occur. If a greenhouse is illuminated with just blue or green lights, then plants growing in that environment will be stunted, then wither and die.

It's often difficult to understand why plants developed this way since green forms a large part of the sun's visible spectrum, but in chlorophyll the pigments are divided into two types, called 'A' and 'B'. Type A absorbs energy mainly from the red and some of the blue photons. Chlorophyll B absorbs more of the blue, but still some of the red photons. These combinations provide the optimum amount of stimulation depending on plant types. Neither A nor B chlorophyll absorbs energy from the green frequency band, hence the appearance of the rejected green light from the plants.

BACTERIA AND MITOCHONDRIA LINKS

The visible wavelengths lengthen through yellow (560–590 nm) and orange (590–630 nm) into the red parts of the spectrum (630–700 nm). Red has great significance for the stimulation of cells and in particular affects the powerhouses of cells called the mitochondria. As previously mentioned, there is an evolutionary link between bacteria and mitochondria, with both originating from the earliest bacteria-forming Precambrian stromatolites. As discussed earlier, my theory is that as more photons enter and are reflected back and forth and absorbed within the bacterial structure, with a build-up of energy beginning to occur to the point of destruction of bacterial DNA. This would naturally inhibit their proliferation. Because bacteria are efficient generators of energy, producing ATP (adenosine triphosphate), theoretically a symbiotic relationship may have taken place between bacteria and developing cellular structures that made use of the energy-producing capability of the bacterium. Bacteria absorbed into cellular structures would be protected by being screened from blue-light photons. Those bacteria therefore evolved to become mitochondria.

Another shared characteristic with bacteria that gives credence to this theoretical relationship between the two is how they reproduce. Binary fission is the method for reproduction in both bacteria and mitochondria. Also, like bacteria, the DNA within them fills much of the internal structure, and because of this there is no mitochondrial or bacterial nucleus. The dimensions are also very similar, and it is this that may be the key to why red-light frequencies have such a stimulative effect on mitochondria, also because of photon resonance. With red-frequency photons having less energy than blue, they avoid mitochondrial DNA damage at the levels of energy penetrating through tissue. These photons slightly raise the temperature by the release of long-wave infrared photons within the DNA structure, which in turn stimulates the production of ATP.

Another explanation along similar lines to that discussed in the previous paragraphs could be that within cells found in superficial layers of the body tissue, penetration of lower energy carried by red photons compared to the blue ones is deeper. Blue is predominately scattered at the surface. Also, some attenuation would occur as the photons filter into the cell. This would suggest that the energy level received by cells is just sufficient to allow a build-up, due to this resonance effect, to a beneficial level for the cell as a whole. If more photons exceed a specific threshold when entering the cell, the mitochondria may be damaged, causing the affected cell to die. This may be just my theory, but it does have some substance in that a researcher (Alexandratou et al. 2002) discovered a detectable 100 Hz oscillation rising out of stimulated mitochondria. It was likened to the energy build-up in a photon-resonant construction found in lasers. This suggests a resonance build-up to and discharge rate of 100 Hz and is in line with my own theories.

RADIANT HEAT

Red light can originate from many sources, such as a fire where the lower temperatures are indicated in the red glowing coals. This is in comparison to the more intense heat being given off, indicated by bluish light coming from the point of combustion. A measure of brightness and colour, using comparisons to natural heat sources, is that of 'heat temperature' of artificial light-emitting devices. The comparisons are a measure of the possible source temperature in kelvin (see Figure 5.7) that are in essence the temperature of radiant heat in the longer infrared wavelengths that generates visible light as a secondary effect.

Lower radiant heat produced from an energy transduction process will begin to glow red. As the temperature increases, the colour changes to shorter wavelengths, eventually causing emission of photons of all the frequencies, which we then observe as white-hot. For specific colours, the heat temperature is the temperature at the source needed to emit the photon of a specific frequency. This simply means that what might be perceived as the 'cooler', but higher-frequency, colours towards the blue end of the spectrum have the highest heat temperature, as blue would be a natural product of higher temperatures in transduction or combustion processes. The reason why blue is incorrectly associated with cooler temperatures may be because of the association with ice looking bluish because of reflection of the blue sky, especially in the polar regions. The temperature of the source emitting photons generates a frequency level at which the photon is emitted. It may be thought of along similar lines to photon energy being directly proportional to temperature of the source that emits it. The higher the frequency, the higher the heat source temperature. Black-body radiation, that is heat given off a perfect black body, would be all frequencies determined by the temperature of that black body.

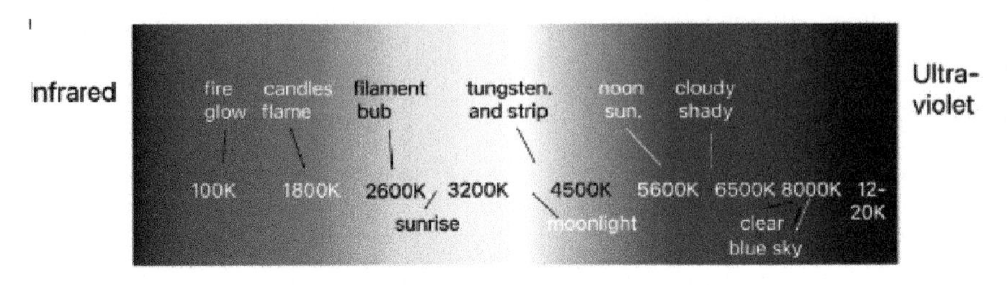

Figure 5.7 Colour temperature scale (kelvin).

The filament of an incandescent bulb may be thought of as a black body, in this case the colour temperature of the light being the same as the temperature of the heated filament causing it to light. All others are just referenced to a heat source. Heat temperature of different colours range from around 1700 K for red to around 27 000 K for blue.

Heated filaments give off red-light wavelengths not only within the visible range but also within the infrared range. The visible light gives the 'warmth' of a pleasing glow but in itself does not provide the heat. The light is the by-product of the filaments heating up in the far-infrared range and further causing visible photons to be emitted along with them. This is due to the increase in activity of those in the infrared range as a result of high temperature. In fact, the hotter the filament gets, the more visible the photons of increasing energies and frequencies that are emitted, until a point arrives where the filament becomes white-hot. The emitted photons then include all those in the visible range. Filament lamps are typical of this, where the tungsten filament rapidly heats up when a current is applied to give off both white light and heat. Most of the energy from a filament bulb is radiated as far-infrared heat. If just required to provide light, then it is a very inefficient use of energy.

PHOTOACOUSTIC IMAGING

A system called 'photoacoustic imaging' (Matsumoto 2018) is being developed at University College London, which is also looking into blood cell reactions when exposed to red-light laser sources. Initial experiments have been applied to the back of the hand. Whilst most red photons pass through tissue, as demonstrated by shining a white light through the nasal ala and the earlobes, the light that passes through is made up of largely unabsorbed red photons, whereas all other frequencies making up the white light are absorbed. However, sufficient red photons impinge on, and are absorbed by, red blood cells, causing a very slight shape-changing reaction. This is termed a 'photoacoustic response' and can be detected as a physical pulse.

If the source of the red light is regularly pulsed, then sensitive transducers can detect these pulses and, if scanned by ultrasonic detection methods, can form an image to show the structure of the vascular formations carrying red blood cells directly under the surface area of where the red light was applied. How the acoustic response occurs is speculative but again may be a resonant effect of photons within similar physical dimensions to those found in the four polypeptide chains that make up a single haemoglobin molecule. The rapid build-up of absorbed energy may change the shape slightly through thermal agitation and, collectively alongside the 260 000 000 others that form a single red blood cell, produce a detectable mechanical pulse.

OTHER CELLULAR REACTIONS

Continuing the idea of resonance, established research suggests that specific frequencies within the red bandwidths optimise reactions within cells. Typical of these is the bandwidth of cyclooxygenase. Karu and Afanas'eva (1995) suggested that frequencies starting around 630 nm wavelengths and at other specific red-light wavelengths gave rise to increased production of nitrous oxide, local to the point of irradiation from a cyclooxygenase reaction within cells. This causes dilation of small blood vessels, increasing blood flow in the very localised area under application. When used in medical situations, this may be beneficial. It may be the result of a photomechanical reaction caused by resonant properties within the cells similar to those of photo-acoustics within red blood cells as mentioned above. Karu and Afanas'eva also showed that the release of prostanoids, which are neurotransmitters responsible for chronic pain reactions, are inhibited by red-light photons accompanied by increased production of ATP, which provides the energy of all bodily actions from stimulus of cellular mitochondria.

DEVELOPMENT OF LASERS

Clear ruby crystals fluoresce at a deep red wavelength just preceding the infrared band. If an ultraviolet source is shined on a clear ruby crystal, the crystal will fluoresce brightly at a pure frequency of 694.3 nm wavelength. This property of ruby crystals gave rise, in the early 1960s, to the invention by Theodore Maiman in 1960 (Townes 2003) of the first laser device. *Laser* stands for 'light amplification by stimulated emission of radiation'.

If a tube of pure and clear synthetic ruby crystal is cut to a specific length, with both ends highly polished, and the crystal is then placed in the centre of a spiral photo discharge tube, such as found in strobes and other flashing devices, then the crystal will momentarily fluoresce with each flash. Photo discharge tubes emit ultraviolet radiation along with a very bright visible white-light flash, and these ultraviolet photons provide the stimulus for fluorescence to take place. The two polished ends of the ruby will internally reflect light back and forth. Many of the photons being produced from the action of ultraviolet photons that cause fluorescence are emitted from the flash tube. Some of these photons will then exit in a straight line at the end of the ruby tube. The photons are caused by electron displacements within the ruby crystal matrix, further causing the generation of photons at a different frequency from the ultraviolet one. If the ends of the tube are silvered, with more on one end than the other, this will increase the build-up to a point that the lesser silvered end will be unable to reflect back all the photons, which then pass through that end as a highly concentrated, parallel beam of red light.

The aforementioned phenomenon is in line with the theories of photon resonance within biological structures that gives rise to the 100 Hz mitochondrial oscillator, along with other, previously discussed effects on bacterial and cellular structures. The similarity comes from the design of this original red ruby laser in that the length of the ruby tube has to be equivalent to a precise, large, multiple of the wavelength of the photons produced by the florescence. The photon resonance then builds up, causing standing waves peaking at each end of the tube. Modern lasers use different materials, including gases, but all have to be designed and constructed with specific dimensions for photon resonance to occur, the lengths being dependent on the colour of the emitted photons.

An aspect of visible laser generated light is that of 'spinkles'. This term, though not often used, describes the effect of reflected laser light that appears to sparkle and twinkle, hence *spinkle*. The reason for this phenomenon is that the generation of laser light causes the emerging beam to be coherent and temporally cohesive. When used in therapeutic lasers, these apply a small but intense beam to tissue as a very bright spot of light. This will be scattered and absorbed in tissue, reducing its temporally cohesive qualities very quickly, but some of it will also be reflected. Because of its intensity, the reflected photons maintain their coherency and are seen as many bright spots of light as part of the backscatter.

USES OF COHERENCE

This reflective coherency property is not just a visual phenomenon but also has many nonmedical uses. It is not restricted to visible wavelengths as it is used in ranging equipment to map areas and to measure the distance from the moon with great accuracy from mirrors placed on the moon during the *Apollo* missions. Because reflections are scattered back in all directions, some of this backscatter will be directed back to the equipment generating the laser beam, where it can be detected. Pulsing the laser and timing how long the reflected photons take to return allows an accurate distance measurement where the transit speed, the speed of light, is a constant. By causing the pulsing beam to physically be scanned up and down, left and right, this allows images to take on a three-dimensional quality, leading to intricate images containing depth information to be reproduced. Lidar (light [or laser] detection and ranging), also known as 'ladar', is a technique that uses visible, nonvisible, and microwave lasers that scan ground contours such as coastlines with great accuracy, providing highly detailed 3D images from the coherent reflections back to their respective sources. This even allows an image through woodland from the air. These are used by aerial archaeologists to locate ancient structures, as sufficient numbers of photons are reflected off the ground through the woodland to form an accurate picture of the ground features.

Laser beams are in themselves invisible regardless of the colour of photons emitted. The spot seen from a laser pointer is the diffused reflection of the spot at a safe level for the eyes to cope with. It is fanciful that sci-fi programmes show laser discharge weapons as powerful straight-line beams, both from handheld 'phasers' and from the spacecraft itself. Laser beams may be visible, to some extent, within the atmosphere, where ionisation of air may cause molecules to emit light when subjected to extremely high-powered energy beams or from the reflection off small airborne particles that would also allow the beam to be seen. In the vacuum of space, unless passing through dust clouds, the beam would be totally invisible.

Holographic images are another feature of the use of coherent fixed-frequency light. If a diffused laser light source is used to illuminate an object and, at the same time, the reflection of it is captured by a photographic film or plate, then an 'interference' pattern will be created and recorded on the plate or photographic film. When developed, this can be used to visually reproduce the scene viewed in full three dimensions when coherent light is shined on the film. The interference pattern when developed will reproduce the scene when subjected to a coherent source of light, with changing angles giving different aspects. No camera lenses are involved. The image is produced because of reflected coherency but can now be achieved by any coherent light source and is not just an effect restricted to lasers. Holographic projection, an idea much liked by science fiction writers, at this time is very difficult to achieve. This is because a reflective medium such as a screen or glass-coated fluorescent sheet would be required and, therefore, only ever be two-dimensional, especially if a live feed is required.

The only possibility for truly three-dimensional images would be to have a scanning and synchronised series of laser beams to form an image in air by ionising the air molecules to produce light at those merging points, where there is sufficient combined energy to cause such ionisation. If a person or object were safely scanned by equally positioned lasers and the detected and processed images fed to those projecting the beams and in full synchronicity, then perhaps a true three-dimensional image in air could become a reality using this technique. This type of use for imaging is not unique in that scanning lasers reproduce images in three dimensions in glass cubes by combining the energy of the beams to melt, or mark, the glass internally to form an image.

Before concluding this discussion on the attributes of lasers, I will say that the properties of lasers are not a natural feature, since lasers are constructed to manipulate photons by their human inventors and do not naturally occur. The exception being the possibility of photon resonance within cellular structures, but not in an intensified form that is emitted as temporally coherent light. However, the frequencies of the photons remain consistent, and their uses, as described above, contribute to the overall purpose of this book. This chapter has just skimmed the surface of visible light interactions in nature. The beauty of the world visible to the human eye arises from photons that are produced within atomic structures. Absorption, reflections, mixing, and emissions produce an almost infinite variety of colours and hues interpreted by the human brain, but we must remember that vision is a construct of the brain as all photons are directly invisible to us. The limit of the longer wavelengths visible to the human eye ends at 700 nm. As we drop lower in frequency, we enter the areas that still affect us, but by other sensory mechanisms within the skin. These are known in the body as nociceptors. They can detect photons from within the next band of lower frequencies, with longer wavelengths called infrared and, therefore, invisible to humans, being outside the visible range of human perception but still visible to some insects and animals.

Chapter 6

FREQUENCIES LOWER —BEYOND THE VISIBLE 700 NM (430 THZ) TO 1 MM (300 GHZ) INFRARED

The infrared spectrum is very important to all living things in that it provides warmth and, in some animals and insects, vision at a wavelength invisible to humans. This, in a similar way to ultrasound, is above the human hearing range and is nondetectable but clearly heard by a variety of animals, notably bats. The lowest frequency of infrared also appears to mark a point in the electromagnetic spectrum where photons cease to have the quality of being particles or specifically defining frequencies. Due to their physical size, atoms cannot emit photons at frequencies lower than around 300 GHz (1 mm wavelength). Photons lose their specific quality of being emitted as individual particles and waveforms. However, within the infrared waveband, both humans and animals begin to detect infrared as warmth and heat through stimulation of specific receptors in the skin. This effect has therapeutic properties similar to visible light used in therapy.

Infrared wavelengths are named in this way because *infra* means below. In this context, *below* means of a lower frequency than visible frequencies as the wavelength extends from 700 nm (430 THz) to 1 mm (300 GHz). This frequency band was discovered using a prism producing a visible light spectrum of the sun's radiation back in 1800.

RADIANT ENERGY LEVELS

Astronomer William Herschel (1738–1822) experimented by measuring the temperature levels of each colour of the spectrum using a thermometer, but then extended his measurement to points beyond the visible red range into the invisible area.

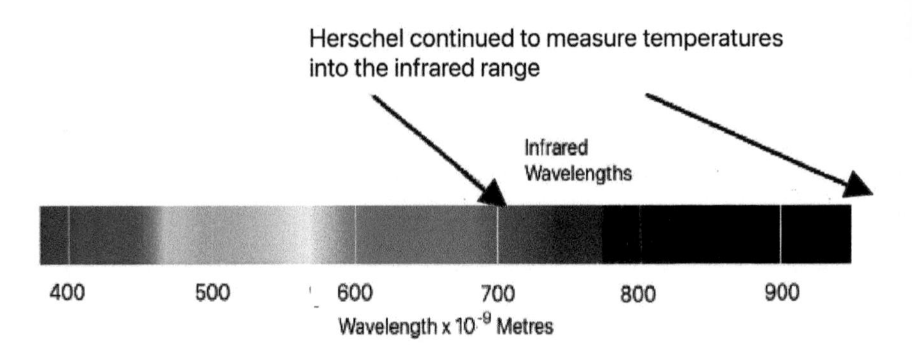

Figure 6.1 Start of infrared band wavelengths where further temperatures were measured.

With further investigation, it was discovered that more than half the radiant energy emitted by the sun were in this region, beyond the visible red, hence called the infrared band. See Figure 6.1. Remembering that the infrared waveband is thought to amount to more than half the sun's electromagnetic output, we can therefore calculate the sun's total energy output along with the total amount of energy arriving at the earth at any one time using the solar constant.

If we view the earth as sitting at the surface of a very large sphere, where the sun sits exactly at its centre, then the solar constant in watts per square metre measured at one astronomical unit (AU) gives a value that is constant for the whole of that sphere's hypothetical surface. See Figure 6.2. The surface area of a sphere is 4^2 where *r* is equal to 1 AU and is the radius of this sphere to the centre of the sun, this being 148.55 million kilometres. This gives:

$$(148.55 \times 10^9)^2 \text{ m equals } r^2,$$

$$\text{therefore } 4\text{JI}r^2 = 2067\text{x}10^{22}\text{m}^2$$

The solar constant multiplied by this number of m^2 shows the total output of the sun to be:

$2.2067 \times 10^{22} \times 1.4$ kW (rounded up)= 3.8822×10^{26} W, or 0.003 8822 yW

For reference, any number followed by 10^{29} in watts is called 'yottawatts'. The established value is 0.00 348 6 yW, but there is variation of about 6.9 per cent between January and July. It is a slightly lower value than my calculated and rounded-up value of 0.003 882 2 yW. More than 50 per cent of this energy is in the infrared band.

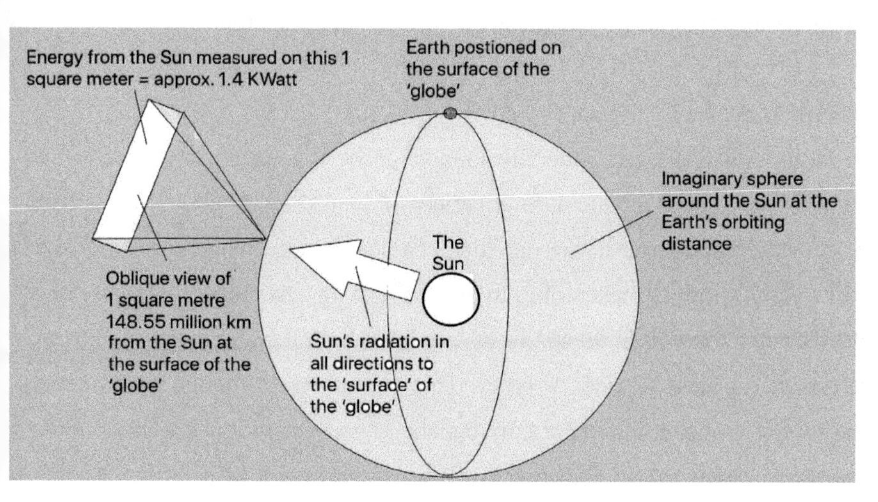

Figure 6.2 Sun's radiation per square metre.

Humans have the ability to sense visible light frequencies using specifically tuned receptors in our eyes. As the frequency approaches 700 nm, the observed red colour gets more intense, then darker, until it can be observed no longer. At this point it enters the infrared band of frequencies at the higher (short wave) end of the infrared spectrum. See Figure 6.1. Starting at 700 nm, other receptors begin to sense infrared as the wavelength extends and the frequency gets lower. Sensory nerve endings called thermoreceptors detect changes in temperature and are part of the exteroceptor group, but in the higher short-wave group, little is sensed under normal conditions.

It is possible to construct lasers in the infrared band, and some are available for use in therapy at around 800–880 nm. Some are at a class IV level and can damage tissue at that intensity if applied without modification. Usually, they are applied in bursts of one billionth (10^{-9}) of a second or could be diffused. Un-diffused and pulsed are possibly akin to wafting your hand across and through a candle flame. If moved fast enough, it causes little or no sensation. If moved too slow, then expect to be burnt. This short duration of application is thought to be therapeutic, but results are largely anecdotal at this time.

THERMAL EFFECTS

Phototherapy is the term generally applied for the practice of using pure visible light frequencies as treatment for a variety of conditions. Some of the research effects were discussed in the previous chapter. At wavelengths starting from just over 700 nm and extending to around 950 nm, therapeutic phototherapy devices still fall within the realm of low-level light therapy (LLLT), with the exception of class III and IV types. This phototherapy, though no longer in the visible range, still is reportedly beneficially effective at these much lower photonic power levels. The energy of the infrared photons is proportionately lower than that of visible light. However, these find their use in many ways other than in therapy, from photography ranging devices to being nocturnal illumination of areas with security or wildlife cameras. The latter are visible only to the camera. Recordings are then electronically converted to be visually displayed as monotone or temperature-referenced colour pictures on screens.

Photons in the long-wave infrared band are also emitted from all objects above temperatures of −268 °C or 5 K (kelvin). All warm-blooded animals, including humans, emit photons in the long-wave infrared band and likewise can detect these photons as heat. Holding a hand close to, but not in contact with, another person gives one the ability to detect these photons by stimulus of the nociceptors (specialised nerve endings in the skin) and is felt as warmth. It is something we take for granted, but it is still the displacement of electrons in the biological molecular formations emitting long-wave infrared photons. So, what displaces these electrons?

Temperature is derived from physical vibrations at all temperatures above absolute zero. This starting point is given as −273 °C or 0 K by graphical extrapolation. This is the point at which all thermal dynamics begin. No thermal activity of molecules exists at or below −273 °C (0 K), so it should not be possible to go lower. However, scientists at the University of Munich (Braun et al. 2013) are carrying out research into atomic activity below −273 °C where, in essence, energy begins to flow from a negative to a positive (colder to hotter) against current laws of entropy stating that heat always transfers from hot to cold. Research is ongoing and at this time discussions on this is beyond the scope of this book. Above absolute zero temperature thermal agitation begins, and this causes displacement of electrons loosely held in outer shells or higher energy levels of atoms. As the atoms regain their stability, the energy of the displacement is given off as a lower-energy photon starting in the long-wave infrared band. As the temperature increases, the number of photons emitted also increases and starts to include photons of higher frequencies extending through the waveband as electron displacements gain sufficient energy to emit photons in the higher-frequency visible band. See Figure 6.3.

With the higher-end infrared wavelengths just beyond the 700 nm visible red end of the spectrum, there is no immediately obvious thermal effect from natural sources that can be sensed by humans, other than a very mild heating sensation if applied to an object or area for extended periods of time. This thermal effect occurs with all electromagnetic radiation that is absorbed and then re-emitted as far-infrared wavelengths. However, electronically generated infrared wavelengths still have their therapeutic uses. Ultraviolet, visible, and infrared photons can be generated by similar devices to match the required

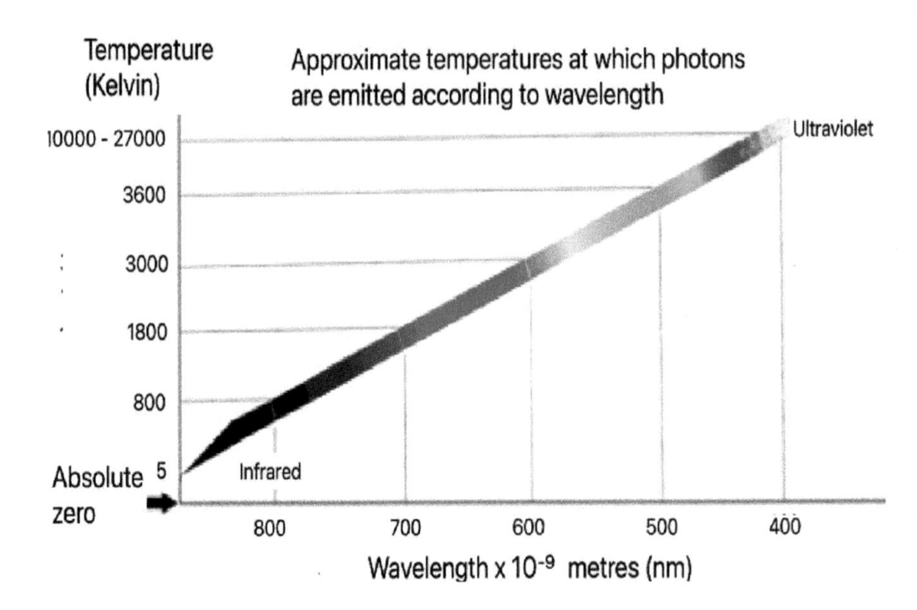

Figure 6.3 Photon emission temperatures.

frequency. As with nearly all photons, danger lies not with the individual photon but depends on the volume applied. The previously listed rating of laser and bright-light devices applies equally to infrared.

LIGHT EMITTING DIODES AND PN JUNCTIONS

Common methods of photon generation are from light-emitting diodes (LEDs). These use layers of specially doped silicon that either adds to or removes electrons from the crystalline structures as a whole. Doping is a process where specific impurities are added during the crystal formation. The two types of crystal are now designated *n* or *p*. N indicates crystals where the matrix contains more electrons because of the type of impurity introduced when the crystal was manufactured, therefore comparatively negatively charged overall. P is where the introduced impurity has fewer electrons and is therefore conversely positively charged. In fact, each piece of doped silicon is still electrically neutral since its balance of electrons and associated protons is maintained; however, 'holes' (electron gaps) and surplus electrons are found within the matrix of *n* and *p* crystals. These are called 'donor' and 'acceptor' crystals, where *n* provides electrons and *p* accepts them. Bonding of n- and p-types together causes a small exchange of electrons across the now formed p–n junction, causing a depletion of holes and electrons layers on each side. This is an area at the junction where electrons cross over into the p-type, creating holes in the n-type that are quickly replaced. See Figure 6.4.

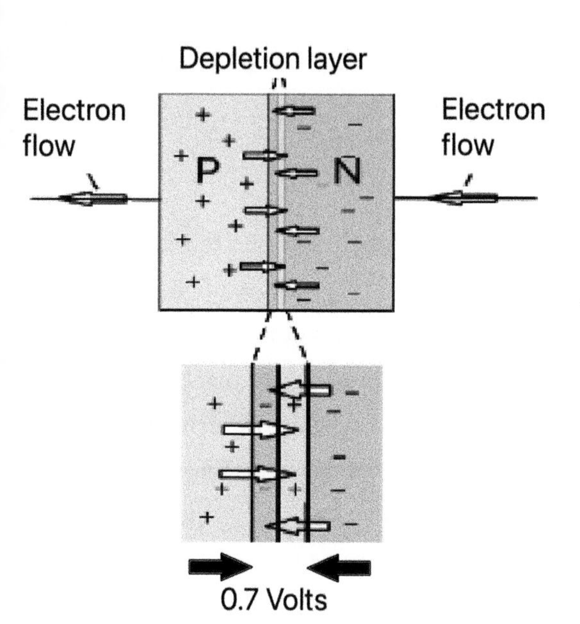

Figure 6.4 P–N junction silicon diode.

This causes a barrier that requires about 0.7 volts to overcome it. Having formerly been an electronics engineer and medical researcher, I find it easy to be diverted and go into great depth as to why certain things occur, but suffice to say that the area along the actual junction forms this depletion layer, and if an electric charge greater than 0.7 volts is applied across the junction, an electric current will flow. This flow will be from the negative supply of electrons connected to the *n* side across the junction to the positive *p* side, but not in the reverse direction if the supply polarity is reversed. Electrons are repulsed by the larger number of electrons in the n-type side of the junction. Since the electrons can then only easily flow in one direction, the structure is designated a 'diode'. The basis of operation of light-emitting diodes is that when the electrons cross the junction through the depletion layer, they recombine with the holes in the p-type layer and, in so doing, release energy as a photon. This is comparable to energy being given off as electrons 'snap' back into position. Different impurities will emit this energy as photons with different amounts of energy and hence frequencies, depending upon the size of the atom receiving them, along with the type of impurities introduced.

THERAPEUTIC USES

Infrared light-emitting diodes are very frequency-specific and, therefore, are of very pure frequencies. Typical of these are wavelengths of 880 nm and 950 nm, but they can produce photons of all frequencies from ultraviolet to infrared and, with modern ones, can be constructed to produce coherent beams as lasers. When they are used as therapeutic devices, the depth to which they penetrate tissue is greater than visible red penetration, but still extending only to around 10 mm at 880 nm wavelengths. At this depth the energy level, that is the number of photons, is reduced by being scattered to around 1 per cent density when compared to the number applied to the skin surface. An increase in intensity does not produce a proportionate increase in the depth of penetration (Turner and Hode 2002).

In my previous books I have discussed electromagnetic therapies and their beneficial effects, including the use of phototherapy. For the purposes of this book, the medical discussion is limited. Higher-frequency infrared photon frequencies are used in many forms of treatment utilising light-emitting diodes (as explained earlier) as simple emitters or specially constructed ones, that is lasers, giving out a straight concentrated beam of light arriving at the surface as a single spot. Unfortunately, the high concentration of photons from lasers causing the photons to be just as quickly scattered in tissue as those from simple non-laser LEDs.

THERMORECEPTORS

As we progress further into the infrared band extending the wavelength beyond one micrometre, we find that thermoreceptors begin to detect infrared photons as heat. This becomes more the case as the wavelength extends to the far infrared. Wavelengths just above visible red are called 'near infrared' and give off very little heat. At the longer wavelengths, nearer to the microwave range, the infrared is sensed by receptors in the skin as warmth and heat. Typical of this type is sunlight. The warmth felt is in the far-infrared frequency range, although both near-range and midrange infrared is also emitted and, like all electromagnetic radiation, is capable of causing some heating if applied for extended periods over an absorbent surface. Both long- and short-wave infrared can damage and burn tissue if too intense. This is especially so as emissions radiated from a fire or flame start as long-wave infrared. The hotter the flame, the more intense the infrared radiation from it, with the radiation starting to emit shorter wavelengths. This includes photons in the visible spectrum, which is seen initially as deep red to orange, and then yellow, and eventually moving to the blue end of the spectrum. When all visible and infrared frequencies are being emitted, this gives rise to white-hot.

PHOTONIC ENTROPY.

Speculating in general about energy going from order to disorder, it is likely that in the future the ultimate transduction of electromagnetism will end up as long-wave infrared photons. This means that all photons of all frequencies will eventually transduce into infrared ones at the lowest frequency possible for a photon, as previously discussed. This is the result of the change from decaying kinetic energy, where this decay will transduce into lower-frequency photons at the longest wavelength at which photons can exist. This is the ultimate and final form of photonic entropy.

In Chapter 2, inflation theory of the universe was discussed, including the idea that there may not be a time when the universe will collapse back into a singularity but will keep expanding. I suggested that all energy making up matter would eventually become cosmic dust on a vast scale. This 'dust' may be scattered infrared photons, eventually broken down into quantum strings after randomly being emitted in all directions as the transduction from higher frequencies to lower ones occurs. They would then be isolated from each other by eventually being so widely dispersed into the infinite void. This may not be the end since inflation theory suggests an energy source outside the realm of the void in which our universe exists. It may be that all the displaced infrared photons or quantum strings are absorbed back into this energy and may once again become part of another Big Bang event.

INFRA-RED DETECTION

Discussion of the infrared spectrum does not end there, as even at the higher-frequency end, the uses and interactions with humans are very much apparent. Sometimes methods of detection are required that allow illumination of areas by emitters found with security cameras, as alluded to earlier in this chapter. A bank of emitters in the high infrared range projects photons at just above the visible range that cannot be seen with the human eye but are visible to the cameras. Reflected infrared photons are focussed on, and stimulate, the electronic receptors, causing electron displacements in the receptor material. These detected infrared photons are electronically transformed into a visual image of the outside area purely illuminated by the infrared. As previously mentioned, there is a very low heat signature with short-wave infrared, so it is ideal for security and ranging as found in cameras. At the other end of the range is the fact discussed earlier that all warm-blooded animals give off photons that can be detected as heat, but also detected by special cameras as found in police helicopters searching for car thieves after they abandon them in the dark of night. An example of a thermal image is given in Figure 6.5, taken by an FLIR (forward-looking infrared) camera. This shows the 'hotspots' on my dog on areas with arthritis.

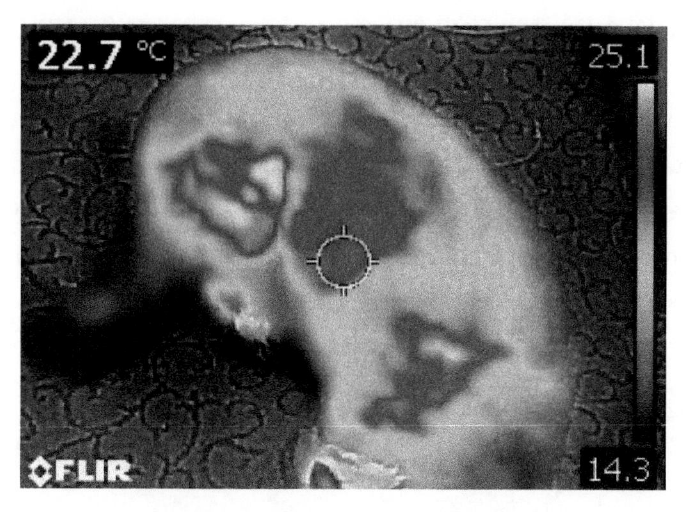

Figure 6.5 Thermal image of my dog.

This chapter has shown the immensity of infrared in the spectrum of electromagnetic energy emitted in the form of photons as a whole. All photons play a part in existence, and the lower frequencies of photons directly interract with life in all its forms. Infrared is probably as essential to us as any frequency in the whole range. The frequencies lower than infrared will be discussed in the following chapter. These are lower than the photonic range but do occur naturally and will be discussed along with artificially generated ones.

Chapter 7

MICROWAVES (300 GHZ TO 300 MHZ)—ORIGINS AND USES

In the previous chapter it was stated that photons do not appear to exist at frequencies below 300GHz. In general, electromagnetic radiation around these frequencies tends to be generated from man-made sources, especially at the lower-frequency end of this band. However, microwaves are found nearly

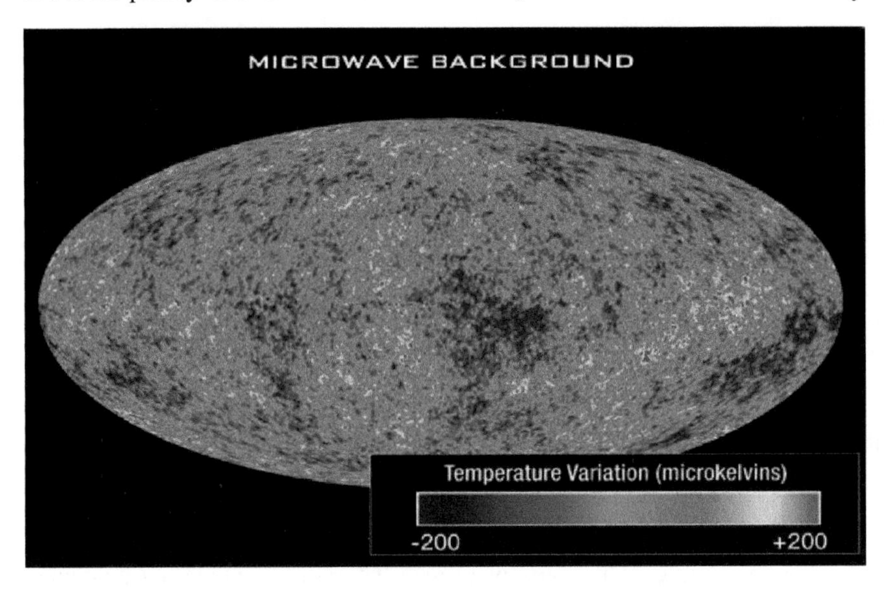

Figure 7.1 NASA's microwave universe image.
This is a NASA image of the early universe looking back 13 billion years. It is a mapping of the microwave signatures from those emitted just after the start of the universe. The image is reproduced here under NASA's free-to-use educational policy. See https://www.nasa.gov/multimedia/guidelines/index.html

everywhere in the cosmos and are thought to be the energetic remnants of the original Big Bang, forming a background noise throughout the universe. As they are not photons in the general sense of being emitted from within atomic structures, they are likely bursts of electromagnetic energy generated at a lower frequency shortly after the Big Bang that traverse and reflect off all material and matter in the universe to form that noise. This suggests that they will eventually spread out, leaving the universe a relatively quieter place. However, the idea that interstellar and intergalactic space is empty is nonsense since the universe is brimming with transiting photons emitted from the most distant extremities in all directions. The NASA photo in Figure 7.1 shows just one frequency band of the many filling the universe with photons.

MICROWAVES

The term *microwave* now is commonly used both as a verb and a noun in that it applies to anything being cooked or heated in a microwave oven and to the oven itself. Microwaves also have similar qualities to light frequencies in that they follow a line-of-sight transmission but are continuous in frequency and not formed of individual photons. Microwaves are a form of electromagnetic radiation of wavelengths between 1 cm and 10 cm (300 GHz to 300 MHz). The study of microwaves as a subject is extensive, and worthy of a book in itself, so all we can do is look at a small part of their function that is familiar because of its being employed in common use. Microwaves form a spectrum at the upper end of the frequency range used in both radar and communication and may be defined by the acronym EHF. See Table 2 for the classification of all electromagnetic frequency acronyms mainly used in man-derived usage, ranging from microwave frequencies to the very longest of electromagnetic wavelengths.

Frequency range (Hz)	Wavelength range	Acronym	Meaning frequency
300–3000	1 mm–100 km	ELF	extremely low
3–30 k	100–10 km	VLF	very low
30–300 k	10–1 km	LF	Low
300–3000 k	1–100 m	MF	Medium
3–30 M	100–10 m	HF	High
30–300 M	10–1 m	VHF	very high
300–3000 M	1 m–10 cm	UHF	Ultrahigh
above 3000 M	< 10 cm	EHF	extremely high

Table 2 Frequency acronyms.

Natural cosmic radiation from stars and the sun includes microwaves emitted in much the same way as higher-energy photons, but is formed by stimulated atoms heated by thermal agitation, including those forming molecular bonds. The rapidly generated high frequency of the thermally agitated molecules causes electromagnetic energy to be radiated as bursts of microwaves, not as individual photons. The whole sky has now been mapped by satellites and ground micrometer telescopic arrays showing microwave radiation from the entire universe. See Figure 7.1. This provides vital information about temperature variations leading to the formation of higher- and lower-temperature clusters.

Most people associated 'microwaves' with specific ovens where microwave energy is generated from man-made sources by methods that have been around for several decades. It is a common misconception by some that the food is somehow changed, these people believing that it stays radiated, almost radioactive in some way. Nothing is further from the truth, as will be discussed in this chapter. The lower-frequency microwaves are possibly different from the more photonic bursts of the cosmic microwaves in that they do form a continuous waveform generated from resonant chambers in devices called magnetrons for as long as the device is activated. Magnetrons are vacuum-tube-like devices that emit electrons from a heated cathode, and these electrons are then attracted by a 2000-volt charge to a circular copper anode that has cavities around it. See Figure 7.2.

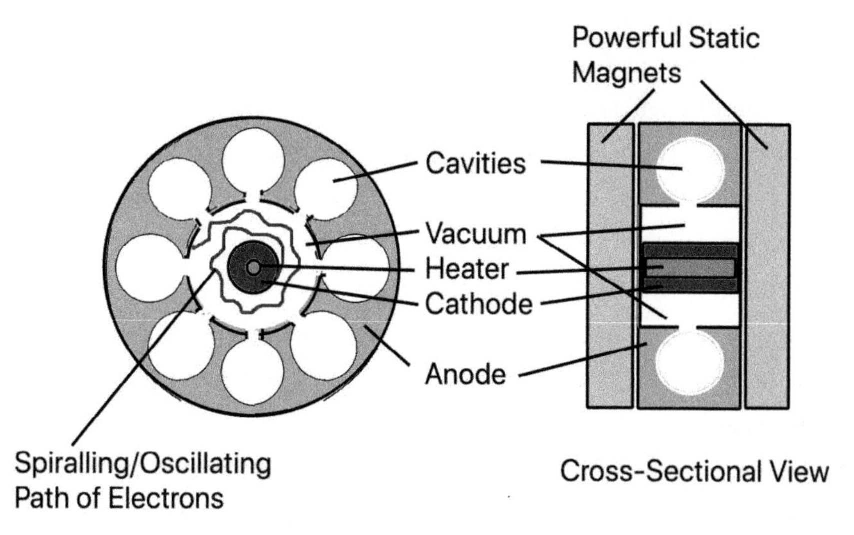

Figure 7.2 Magnetron microwave generator.

Two powerful and complementary field magnets provide a strong aligned and static magnetic field across the structure. This causes the electrons attracted from a heated cathode by the 2000-volt anode potential to spiral in the vacuum, passing over the cavities and within the magnetic field. This is called a 'cyclotron effect'. A resonance is set up in cavities by the flow of electrons over each of the cavity orifices, affecting the spiralling electron flow until it is eventually absorbed by the anode, causing an oscillation in the electron flow at a microwave frequency. This frequency is at the designed and determined frequency set by the resonant cavities, providing a high-intensity signal that is then fed by a waveguide into the oven.

Water molecules are weakly diamagnetic, and as the microwave electromagnetic waves enter the oven, the molecules in anything containing water try to align with the microwave frequency field. In doing so, they flip back and forth as they attempt to line up with each individual half of the microwave cycle. This causes a physical ultrahigh-frequency vibration of the molecules that then generates heat through the emission of long-wave infrared photons sufficient to cook any food containing radiated water molecules. The frequency used to cause this effect is 2.45 GHz. This is not the resonant frequency of water molecules as I have heard many people suggest. If such a resonant frequency were possible, it could break the molecular bonds of H_2O, releasing hydrogen and oxygen separately, which, as gases, are a dangerous combination in a closed environment. Other diatomic molecules will also vibrate and heat up as well if radiated with microwaves. The cavities in the design of the magnetron anode are another example of resonant chambers similar to that within cells discussed in earlier chapters.

The energy of these artificially generated microwaves is no longer present in individual packages, as with photons, but changes to be a continuous waveform of energy like a sort of continuous photon, without being a particle as previously discussed. There are many aspects of it that still possess some of the characteristics of individual photons, the main one being that microwaves, like all electromagnetic energy, radiate at the speed of light. Here the close link between magnetic fields and photons begins to emerge. Photons come from shorter-duration electron transits within atomic structures. The displacement energy is radiated away as a specific package of energy made up of a frequency related to the energy of the displacement. A magnetic field is produced by an electric charge pushing electrons in continuous transit to flow between atoms in an orderly direction, in the process forming the magnetic field at right angles to the flow. Like a photon, it is not specifically electrically charged. However, if the field is made to vary, it will cause electrons in other materials to form oscillating charges if they are placed within that field. This is termed 'mutual induction'. A flow of electrons that is fixed and constant will produce a static magnetic field which in its own right cannot cause induction. As stated in Faraday's law, 'An electric charge can only occur in a time varying magnetic field.' Noninductive magnetic fields are from within the structure of static or permanent magnets, although movement of an object within such fields can cause some induction as long as the relationship between field and object is one of relative motion. A good example of this is a coil wrapped around a rotating arm in a static field, which will cause a charge across the winding whilst rotating. This is the principle of an electric generator or, more commonly named, an alternator.

Microwave generators, in common with lower-frequency oscillators, cause the electrons to go back and forth in the conductor. This produces a corresponding magnetic field that follows suit by rising and falling in intensity. This effectively becomes like a continuous photon with a specific frequency that, also like a photon being emitted, is now radiated. The continuous property of generated microwaves can be turned into short-duration pulses, used in many situations such as radar and communication. Microwaves can be transmitted as a continuous waveform that can cause heating when using the higher frequencies, but it can also be of different intensities, otherwise known as amplitude.

Another characteristic is that microwaves can be focussed like light and can be used to transfer energy. Because of its high frequency, the microwave 'radiates' from aerials or antennas to which the microwave generators are fed. Like all electrons oscillating back and forth at a microwave frequency rate forming oscillating charges, an oscillating magnetic field is produced at right angles to the electron movement and, with specifically designed aerials, produces magnetic fields around, or directed from, the aerial assembly. This field is also changing at the microwave frequency rate. At such high frequencies, a phenomenon occurs where the expansion of the field is radiated away from the aerial at the speed of light. Since the speed of light is the limiting factor, this electromagnetic energy leaves its source in all instances at this ultimate speed, exactly the same as photonic emissions from atoms. It is therefore said to be 'radiated', partly because when using a single aerial element, it is omnidirectionally, or radially directionally, transmitted, but with microwaves the transmitted fields are by design caused to be more focussed.

Under controlled conditions, microwave energy has to be confined as it is fed to an 'antenna' by conductors specifically designed to optimise transmission along the antenna and to reduce energy loss through radiation along its length. This is achieved at such high frequencies by using 'wave-guides'. These resemble rectangular cross-sectional pipes that are designed in their dimensions to match the wavelength of the microwaves. The effect is to confine the microwaves internally within the waveguide as microwave-frequency electric fields are formed within and along the structure. The antenna is a device that also matches its dimensions to the frequency wavelength and is therefore 'tuned' to the frequency that is being transmitted. The antenna acts as an electrical field to the electromagnetic energy transducer. It is similar to the radiating antenna in microwave ovens as discussed earlier.

USING MICROWAVES FOR DATA TRANSMISSION

In the case where microwaves are used for data transmission or radar, they are directional in that the radiation can be focussed. Microwaves are generated as a signal by a magnetron and again become electromagnetic in nature when radiated, but this time over long distances. As they are electrical wave formations, they are transferred to the antenna by longer waveguides. These are also rectangular tubes, as discussed above.

At the start of *Ripples in the Ether* it was stated that for electromagnetic waves, a transmission medium or 'ether' is not needed and does not exist. However, in transferring microwaves to an antenna, a method of containment for this energy is required, which correctly dimensioned waveguides provide. From tuned cavities, the magnetron creates electric pressure waves by the variation of electrons at microwave frequencies directly from magnetron generators. These cause the release of electrons from atoms in the conducting medium. These electrons are quickly replaced, but the action effectively causes a ripple along the connected waveguide that contains the electric field. This is by the movement of electrons in and out of atoms along it. At the antenna, the oscillating waves then allow this photon-like electromagnetic energy to be released and radiated away at the speed of light, as discussed. Since this radiation does not have temporal cohesion, some form of focussing is required to target an area or a receiver antenna some distance away. This is usually in the form of a parabolic reflector.

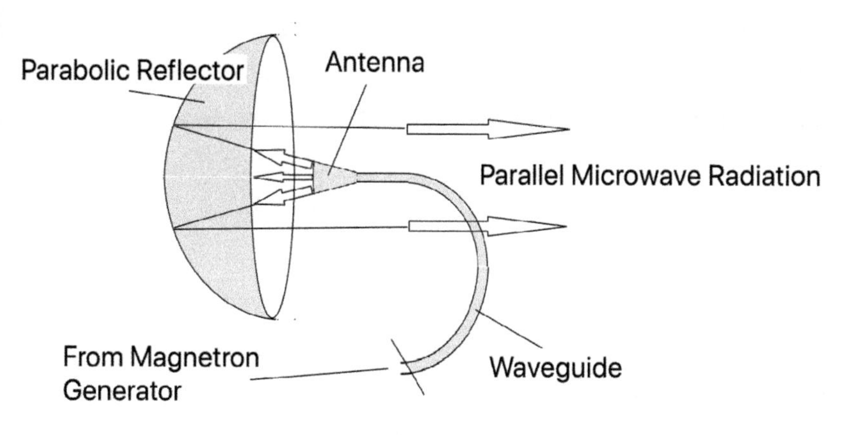

Figure 7.3 Parabolic microwave antenna.

Figure 7.3 is drawn for simplicity to illustrate this, but it serves to demonstrate the principle of divergent microwaves being directed into a straight beam, much like a focussed torch. With microwaves used in radar, they are pulsed and reflected from distant objects. The parabolic reflector receives some of those reflected waves and refocuses them back to the antenna, which then works as a receiving antenna. Timing and angles of the microwave beam then provide the position and range of the object reflecting the waves.

With very focussed microwaves, there was a suggestion that they could be used to transfer energy from a source to a receiver at levels sufficient to be of use to power equipment and lights and to feed into the domestic supply. In the 1970s ideas were put forward for solar energy collectors in orbit that would capture sufficient amounts of energy to be converted into microwaves on board the satellite and transmitted with focussed antennae aimed at special receiving stations on earth. I have looked for further reports about this, but there is very little information around to suggest that it was ever seriously attempted. Logically, if the focussed microwave energy were at industrial levels, there could have been catastrophic consequences for any object or aircraft electronics accidently interrupting the beam. However, back in 2013 it was still on a NASA wish list to develop the technology.

The microwave range covers a wide spectrum, including extremely high frequency (EHF) and ultrahigh frequency (UHF). The range is commonly associated with microwave ovens, as previously discussed. There are specific categories covering each band within the general microwave range, classified as S, C, X, K_u, K, or K_a. However, since *Ripples in the Ether* covers the generalities rather than specifics, going into the reasoning for these divisions is not part of the book's remit, since deep discussions would be of use only to those involved in studying radar design and operations. As we progress further down the frequency range, we are still in the microwave band, but at the lower end, now described as EHF or UHF as above.

In summary, this chapter has shown that there is little difference between dynamic magnetic fields and electromagnetic photon radiation. Both magnetic fields and photons are a product of electrons being active both in entering and exiting atoms and being forced to change energy levels within atomic structures. Photons, once emitted, will carry on forever, never losing their individual cohesive integrity unless encountering or being absorbed into other atomic structures. Magnetic fields can be either static as a result of an even drift from atom to atom inside a crystal structure, or more dynamic, where they can be made to do work such as that described earlier with magnetron generators. Both photon and dynamic magnetic fields are able to transfer energy at levels determined by their frequencies, so they are in essence different manifestations of the same energy form, varying only in frequency and intensity.

Chapter 8

COMMUNICATION FREQUENCY WAVES (UHF) 3 GHZ TO < 20 KHZ

The ubiquitous cell towers that have sprung up all over the country are evidence of the use of microwaves in the lives all mobile phone owners. Each mobile phone is in itself a microwave generator and receiver, allowing two-way communication with those towers. Generating microwaves from magnetrons was discussed in the previous chapter, as was the use of microwaves in radar and other high-power systems. In other cases, as with mobile phones, they are generated by the use of quartz crystal oscillators built within the phone's electronics and used at very much lower power levels. These types of oscillators are used throughout the radio frequency range alongside the older, established inductive-capacitive types. This chapter takes a look at how this wide range of frequencies is applied.

USE FOR CELLULAR PHONES

As has been discussed in the previous chapter, microwaves can be used for communication. Cellular telephones use frequencies from around 800 MHz to 2.6 GHz, where voice and other data is coded uniquely for each phone. The technology involved uses the high frequencies as 'carrier' waves, where information is digital and changes the otherwise continuous waveform into coded bursts in sequences of 1s and 0s. This is achieved by rapidly pulsing the waveform in specific patterns by effectively switching it on and off in coded sequences. It also involves 'cells' where most areas of the country are covered by a matrix of aerials that have mobile phone signals induced into them. See Figure 8.1. A central computer identifies the aerial with the strongest signal from all those receiving an individual cell phone's transmission.

Figure 8.1 Remote microwave communications station.

That cell is then allocated a frequency through which audio and data are transmitted and received. If the cell phone is travelling, then the system switches between cells to maintain the best contact with the phone. All this allocation and selection is carried out almost instantaneously, and each individual phone is identified by its own unique identifier code.

At this point some readers may have noted that cell phones work in the microwave range and, after reading the previous chapter, may be concerned about their effect when used in close proximity to the head or the effect of placing aerial towers on or close to schools and housing. It is a genuine concern, but if looked at logically, we see many differences between microwaves used in an oven and radar and those emitted from a cell phone. These differences are as follows:

a) Power levels

The power levels in microwave ovens are in the range of 750–1200 watts or more. The output power of a cell phone is three watts or less. The amount of energy capable of causing heating would be down to far less than one watt total energy density in the direction of the head and would spread out because of the omnidirectional radiation from the phone's antenna. This follows the inverse square law of attenuation and would mean that energy density concentrations at any specific point would be in the range of milliwatts.

b) Microwave energy

In an oven, microwave energy is concentrated, continuous, and reflected within the chamber until absorbed by food. Microwaves used in communications are interrupted by digital modulation and so are never continuous and are of a much lower average power level in terms of direct power-for-power comparison.

c) Microwave oven energy levels

These could cause burns to tissue through heat over a short period of exposure if such exposure were possible. It is prevented by a locking system that stops the generation of microwave energy if the oven door is open. Cell phone power levels may cause extremely minute heating of water molecules within the head if held very close by the phone's own antenna and exposed for long periods during communication, but there are conflicting views on whether or not this is detrimental at what would be just a very slight, and almost immeasurable, increase in temperature.

d) Microwave frequencies
These are not capable of ionising molecular structures compared with higher-frequency and energetic gamma ray, X-ray, and UV photons, and because of this they logically cannot be considered to be cancer-forming (carcinogenic).

e) Transmitted microwave energy from mobile phone aerial towers
This is usually radiated omnidirectionally and therefore diminishes in energy density very rapidly following the inverse square law. The amount of energy available to affect humans in the vicinity of the transmission tower is quickly down to microwatt levels induction into tissue, this after a very short distance from the tower. Heating of radiated tissue would occur only if in contact with or within a few centimetres of an antenna, and only if the antenna is in active transmission. If contact were possible, radiation would give rise to radio frequency burns the same as those from any exposure to heat from any source. However, the chances of getting close to any antenna is highly unlikely as the antennas are both encased and usually placed high up and out of normal reach.

It is my personal view that there is more likely to be a greater localised subdermal temperature increase when the head is directly exposed to sunlight. Any minute temperature increase would usually be taken care of by the body's homeostatic response to balance out any unnatural temperature change, as happens in any part of the body. It is also my view that as only water molecules are likely to be very slightly affected, then tissue structural damage caused by heating is highly unlikely ever to occur at such minute levels.

The foregoing explanation of cell phone networks is very much simplified but serves to show how ultrahigh frequencies can be manipulated to provide effective communications to millions of mobile phones simultaneously. It may be noted that the overlap between microwaves and UHF ranges is at 3 GHz.

Starting at slightly longer wavelengths of this, EHF, the lower-frequency range, is used in communication, not just with mobile phones, where only a short range is required, but also in situations where ranges up to several tens of miles are required, particularly in aviation communication. With these high frequencies, a line-of-sight requirement limits the range. For aircraft, maintaining contact with the base or airport requires that it be within visual range. This is not in a literal sense, but at height to the visual horizon that could be forty-nine to fifty miles away, dependent upon aircraft altitude. This makes it more effective over greater ranges at the higher flight levels both from air to ground and, conversely, from ground to air. It is also used in local area control. Sometimes repeater stations situated on hilltops extend the range. Police communication systems have repeaters that allow communication with their base to and from ground levels and within valleys.

In the early stages of radio voice communication development, the method

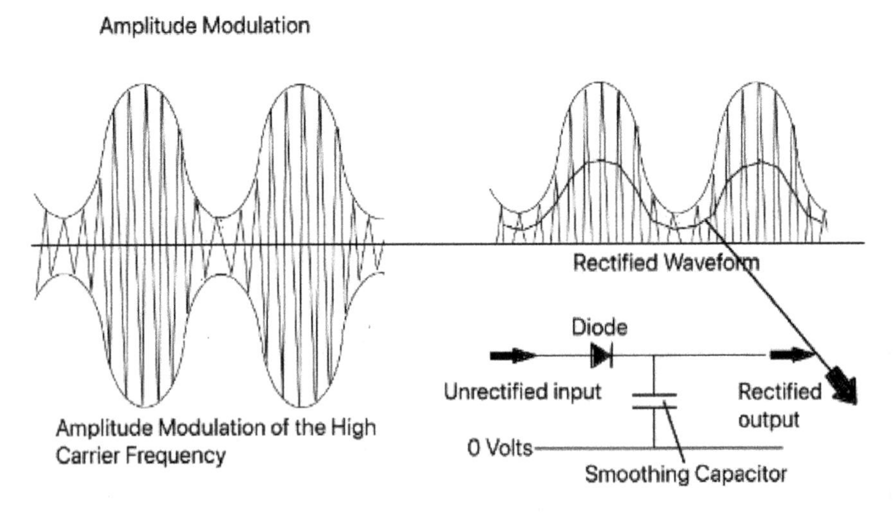

Figure 8.2 Amplitude modulation.

of encoding speech was through modulating the amplitude of the UHF, VHF, and HF waveforms. These are still referred to as carrier waves. This method is still in use today and is known as AM (amplitude modulation) transmission. The modulation is a variation of the positive-to-negative amplitudes of the carrier wave. These are referenced about a central zero point and vary in line with an impressed audio signal. See Figure 8.2.

The carrier frequency is generated by specific circuits called oscillators. The components making up these oscillators were initially made up of coils and capacitors in what are known as L–C resonant circuits. For those readers not conversant with electronics, the L represents an inductance value in 'henries', made up from a coil of wire of specific dimensions. C is the amount of charge that a capacitor can hold, measured in farads.

PRINCIPLES OF THE COIL

A property of a coil is that it builds up a magnetic field both through its centre and around itself when a current is passed through the wire forming the coil. This is stored energy that collapses when the current is switched off and, in so doing, induces a voltage across itself with the opposite polarity to that which formed it as the field collapses. This is referred to as self-inductance, creating an induced 'back electro motive force' (BEMF). It can be very high but is moderated by other components such as a capacitor, as mentioned above, which charges up when connected across the coil. It then discharges back through the coil, forming another magnetic field that will then collapse, so repeating the process and again charging the capacitor. This is an electronic pendulum, and like a free-swinging pendulum, it will decay as it loses energy with each cycle as the charged capacitor will then discharge through the coil, repeating the process. In electronics, by designing the L–C oscillator circuit to allow a little positive feedback to the coil and capacitor enables the charge–discharge cycles to be constantly maintained.

USE OF QUARTZ

The use of quartz eventually took over from pure L–C circuits by using a property of quartz called the piezoelectric effect. In essence this is a property of the crystal structure of quartz that, if compressed, causes a momentary electric charge to develop across the crystal plane during the dynamic time of compression. If the pressure reaches a maximum, the charge balances out, returning to zero. The opposite occurs when the pressure is removed, momentarily causing a charge of the opposite polarity.

Since all solids have a mechanical resonant frequency, quartz can be laser cut to any size with a resonance specifically chosen that matches the desired frequency. If a voltage is applied across the face of the crystal, the charge causes the crystal to momentarily deform. By detecting the voltages generated as the quartz deforms and recovers, and then positively feeding back the voltages into the crystal circuit, these changes are reinforced, sustaining the vibration and resulting in a stable waveform. We then have the basis of a crystal oscillator. This gives out a very precise and stable signal that can be used to form the carrier wave for communication. These crystal oscillators are used extensively now in watches and other timing devices, especially in computers where the clock speed can be many GHz.

MODULATION

Modulation, the method of encoding speech on a carrier wave, is achieved through varying the amplification of the generated frequency and combining it with a speech signal from an audio circuit as discussed earlier. Another type of modulation is to take the carrier wave and cause it to slightly vary about its mean frequency at a rate carrying the audio signal. This is known as FM, frequency modulation. These two methods remain in use today, but with the extremely high frequencies now available, digital encoding provides the best use of these high-frequency properties. This allows many channels specifically coded on the same carrier frequency to operate simultaneously, which is called 'multiplexing' and requires methods achieving absolute synchronisation between transmitter and receivers.

CARRIER FREQUENCIES

If we take the case of a single connecting line, such as found with telephone landlines, then we find that multiplexing can have many inputs. An audio signal can be electronically sliced into many sections of short duration related to a specific synchronised allocation slot. This can be nanoseconds per segment with up to 20 000 parts per second per channel. Many millions of segments from other audio inputs interlace together within that second. Since each segment is time-related, the receiver electronics will only allow the signal from a single input to be accessed within each specified time slot. The signal, if simply an amplitude modulated one, is then electronically reconstructed to reproduce the original speech or music patterns. If the signal is digitally encoded, where binary values are assigned to audio levels, then the same principle would apply, and a circuit called an analogue-to-digital converter is employed. Using light instead of EHF as the carrier medium allows a far greater number of channels to operate through fibre optic cables.

Ultra-high carrier frequencies using microwaves can be operated in the same way as described above. The transmission with EHF and UHF communications is similar to the transmission light. It is ideally a straight line from transmitter to receiver as previously discussed. Television transmitters are usually mounted on very tall towers or atop hills overlooking the reception area. Where houses are not in the visual line of sight to the transmitter, repeater stations receive and retransmit to lower areas out of the line of sight from the main transmitter. Receiving aerials have to be tuned to the frequency of the transmitter. The link between electromagnetism and radio transmissions is apparent in that the transmitter sends out its transmissions at very high-power levels, usually omnidirectionally. The transmitted energy follows the inverse square law, which means that within a very short distance the signal level weakens substantially, usually for every unit distance, as compared to the amount of energy supplied to the transmitter aerial.

The signal received (induced) in the receiving aerials at some distance away is measured in microvolts (with a microvolt being one millionth of a volt). Electromagnetic energy even at this very low level causes the electrons in the aerial metal to be slightly displaced by this microvolt induction. If the length of the aerial matches the wavelength of the frequency being received, then a resonance will take place, allowing an increase in the induced voltage. The strength of the signal can further be improved by a system of added aerial elements called reflectors and directors. This gives the familiar shape of television aerials currently seen on houses with terrestrial (not satellite) signal televisions. These aerials have to point in the direction of the transmitter and be placed either horizontally or vertically in orientation to match the orientation of the transmitter aerial in order to be of maximum effect. The frequencies currently used for terrestrial television are 470–884 MHz.

Satellite television makes good use of the line-of-sight capabilities of ultrahigh-frequency transmissions. The ubiquitous satellite dish receivers point directly at the geostationary orbiting satellites that are, in effect, repeater stations for transmissions originating on the earth. Microwaves are the key to these systems as they are able to be focussed directly on the satellite-borne receivers in geostationary orbit at around 20 000 miles above the earth. Encoding of TV and other communications is digital in ways previously described, allowing many hundreds of channels to use one particular waveband.

Another use of microwaves is in global positioning systems. Strategically placed satellites orbit high above the earth in fixed orbits referenced to each other and transmitting signals. These signals are picked up by satellite navigation (satnav) GPS receivers using multi-channels. Since the timing of these satellites is known precisely, then with just three being visible at one time, it is possible to start to identify the position on earth of the receiving unit by triangulation methods according to the position of the receiver. There are thirteen receivers in each unit. As more satellites become visible to the receiver's algorithms programmed in to the equipment, they are able to ground-position the receiver to within a couple of metres. There are many other uses of satellite microwave communications, including worldwide telephony, which allows phone calls to all regions of the world by the use of satellites and both ground- and space-borne repeater stations.

This excursion into ultrahigh frequencies gives a flavour of the methods of producing them and a sense of how they can be applied in both a heating sense and a telecommunication sense. The very small voltages induced into receiving aerials are electronically amplified to useful levels, especially for those radio transmissions that are amplitude modulated (AM). This impresses the audio, speech, or music waveform on the power levels, transmitting the carrier waveform.

Recovery of the audio component is achieved by the rectification process. This is required because if an unrectified signal in its purest form were to be passed straight to a speaker, then nothing would be heard. The speaker could not possibly respond to the high-frequency carrier waveform. The average level of the audio signal would be zero, with each half of the waveform cancelling the other out. A simple way of getting around this is to cut the modulated waveform in half by passing it through a one-way electrical device called a diode. See Figure 8.2. If it were then fed to a speaker after this rectification process, the impressed sound would be discernible because the speaker, unable to repeat the very high-frequency carrier, would only respond to the average of the remaining rectified one. Hence the modulating sound signal would then be heard, and the high-frequency component smoothed or filtered away by the capacitor as in the simplified circuit illustrated in Figure 8.2. Of course, this is a very simple explanation, as other things can interfere with the received signal and any spark or spike caused by switches and so forth may be heard as a crackling or spitting sound.

FREQUENCY MODULATION

The other form of common modulation is to cause the frequency to vary (frequency modulation (FM)). See Figure 8.3. Receivers tune in to the mean carrier frequency with highly selective tuned circuits. As the frequency varies, the selectivity of the discriminating circuits sees that the frequency gain varies at the rate of the impressed modulation. This is then utilised and turned into an audio signal. The advantage here is that anything that causes a jump in amplitude, such as the noises generated as discussed with AM receivers, is greatly reduced as it does not affect the frequency variation.

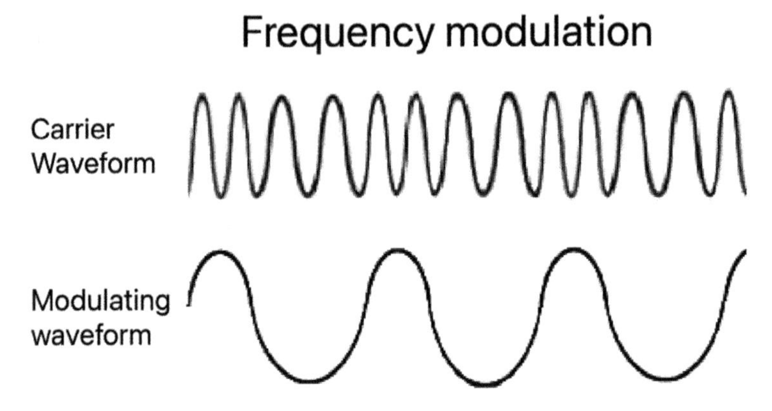

Figure 8.3 Frequency modulation (FM).

There are two main problems with simple radio receivers, firstly adjacent channel interference (nearby channels very close to the desired frequency) and secondly second-channel interference (frequencies that are harmonics of the desired frequency, e.g., 2× and 4×). This is overcome by making the receiver more selective by using a process known as heterodyning. To tune in a receiver to a specific frequency requires that one of the components of the tuning circuit be variable so that its resonant frequency matches the desired channel frequency. This was achieved by making the coil fixed and designing the capacitor to be variable. The formula $1 / (2\pi\sqrt{LC})$ calculates the resonant frequency of the tuned circuit.

Once the selected frequency has been tuned to a specific frequency and amplified, an oscillator, whose frequency is coupled to another part of the same tuning capacitor, provides a frequency that is always a set value below that which is desired. Both the original and the local oscillator frequencies are then added together to produce a harmonic waveform. This causes 'heterodyning' and has several components. These are the original frequency plus the local frequency; original frequency minus the local frequency; and difference frequency. This difference frequency, which will always be fixed, is called the intermediate frequency (IF) and is then selected by another tuned circuit that can then be rectified. This method removes most of the interference from either second- or adjacent-channel interference. The process may then be repeated, incorporating fixed tuning components to provide further selectivity in what is called a 'double het receiver'. This further improves other channel interference rejection.

As communication frequencies become lower, the range becomes greater as the signals begin to follow the curvature of the earth by reflecting off the ionosphere. Amateur radio enthusiasts use HF to communicate directly around the globe with other hobbyists.

The lower frequencies used in radio transmissions are down to around 20 KHz. The wavelength of a 20 KHz waveform is 15 000 m or 15 km. They are more in line with expanding and collapsing magnetic fields than with radiation. They are used to communicate around the globe with marine and submarine vessels; however, these are not the lowest radio frequencies as they go down to between 80 Hz and 40 Hz. The antennae for such frequencies are extremely long. Usage and operations are classified, but these low frequencies (50 Hz and 60 Hz) are probably used to communicate with submarines at extreme depths. They effectively become like the primary winding of an enormous transformer, with the receiving vessel-borne antenna acting as the secondary wherever in the world's oceans it is to be found.

DOMESTIC AND INDUSTRIAL ELECTRICAL ENERGY DISTRIBUTION FREQUENCIES

This chapter has so far focussed on communications, discussing a wide spectrum of frequencies. However, there are one or two much lower frequencies that are of great importance and that directly affect virtually every person alive. I include them here since they provide the energy for most of the ground-based communication systems. These frequencies, 50 Hz and 60 Hz, are used to provide electrical energy from generators to users. These frequencies are important in that they allow the efficient transmission of electrical energy. The users concerned include industry, commerce, and domestic, this last being home use. In the UK and many other countries, 50 Hz is used as the preferred frequency, but the USA works on a frequency of 60 Hz. These very low frequencies are because of the spinning speeds of the generator rotors, which are 3000 and 3600 revolutions per minute respectively. These spin speeds are likely to be the limiting factor because of the size and methods employed to convert heat or wind into rotational motion and then into electrical energy. Steam, diesel, nuclear, and wind are the common power sources currently employed.

Electrical energy is transferred via grid networks from source to users at very high voltages as sine waves in three phases at 120-degree differences. The generators rotate synchronously throughout the land, with each power station providing an input into the grid. The peak voltage, around 6000 volts RMS, is usually provided directly from the generator at large power stations. This is stepped up by transformers to very high voltages of up to and beyond 400 000 volts and is distributed by the ubiquitous pylons covering the entire country. The main 'super-grid' carries the very high voltage, with sub-grids stepping down to 125 000 volts. For domestic supply, the voltage is further reduced to 11 000 volts to substations, where there is a final reduction to UK domestic supply of 220 volts to 240 volts. In the USA it is lower at 110 volts to 120 volts.

In my days of teaching electrical and electronic engineering, questions that have been put to me about the domestic electrical supplies have included the following:

1. Why do we use alternating current instead of direct current?
2. Why are such high and dangerous voltages needed to distribute electricity?

The answer to the first question is that in the early days of electrical supply, very few areas were located close enough to a generator and those that were, in fact, were supplied with direct current. The voltages were taken from the generator and rectified to give a single polarity. This rectification was probably achieved by the way current was accessed, by the commutators of the generator rotor spinning the coils in a strong static magnetic field. These were designed and connected to provide a pulsed DC voltage that could then be smoothed with capacitors. This was fed to the consumers directly at this generated voltage but would have been very limited by the distance in its ability to deliver a sufficiently high current to satisfy all the needs as more loads came online to be electricity driven.

In 1882 the first electrical supplies were provided in London by the Edison Electric Light Station for street lighting at a voltage of 110 volts DC. This supply ran at a loss, possibly because of the reasons given in answer to question 2, which follow: In 1891 the Deptford Power Station was built and run by the London Electric Supply Corporation. This provided an AC single-phase output, based on a design by Sebastian Ziani de Ferranti (1864–1930), a British electrical engineer, and increasingly developed by industry.

In the USA, Nikola Tesla (1856–1943) designed similar systems based on alternating current generation and the polyphase system that we now know as three-phase supplies. As the demand for electricity increased by more households coming online, it was soon apparent that the direct current method of supply was very inefficient and incapable of meeting the ever-increasing demand. This led to a larger, centralised, more powerful generator where the current was accessed directly from commutator rings in an unrectified form as a sine wave.

The frequency of the sine wave was standardised in the UK at 50 Hz by keeping the rotor spinning at 3000 revolutions per minute. This AC voltage could then be stepped up to a very much higher voltage by the use of transformers and stepped down again to the voltage required at the end user's domestic supply or supplied at a higher level for industrial use.

The answer to question 2 is straightforward in that the higher the voltage, the lower the current and energy losses in transmission. Since power in watts = voltage × current ($P = V \times I$), if a small town consumed, say, three megawatts (3 000 000 watts) and the electricity was supplied from a generator to the town at the domestic supply of 240 volts AC, then the current required would be $I = P / V$.

This would come to:

$$3\,000\,000 / 240 = 12\,500 \text{ amps}$$

Since there are no zero-resistance transmission cables, and even if the resistance to just one small town was at an incredibly low resistance of 5 ohms, the power lost in transmission at 240 volts would be as follows:

$$I^2R = P = 156\,250\,000 \times 5 = 781\,250\,000 \text{ watts}$$

This is more than the output of a large power station lost as heat in the transmission lines alone. By using much higher voltages, the heat loss in transmission is greatly reduced. Taking the above-mentioned example of 3 000 000 watts required, when using the same method with a grid supply of 240 000 volts, the transmission current would then be:

$$I = P / V = 3\,000\,000 / 240\,000 = 12.5 \text{ amps}$$

The heat loss in transmission is now:

$$I^2R = P = 12.5^2 \times 5 = 781.25 \text{ watts}$$

This is less than a single-bar electric heater. It is probable that power losses are higher since I have based my calculations on a transmission resistance of 5 ohms. It may be higher than this; various sources suggest between 0.1 per cent and 4 per cent transmission losses overall. The simple rule is, the higher the voltage, the lower the losses—and transmitting each phase on different lines reduces further heat losses. Figure 8.4 illustrates a typical high-voltage pylon.

Figure 8.4 Typical three-phase high-voltage pylon. The lines on top are earth reference lines.

Domestic supply voltages are given as 240 V RMS, with RMS standing for root mean squared. It is stated to show an equivalence of a direct current (DC) voltage having the same heating effect through a resistance wire as the alternating current. The peak value of a 240-volt AC supply is 1.414 multiplied by this value. The 240-volt RMS is actually 339.36 volts peak voltage. Peak positive to peak negative gives a variation of 678.72 volts in UK household domestic supplies. AC voltage waveforms have a sine wave shape and a frequency of 50 Hz. In the UK, this is due to the generators all rotating synchronously at 3000 revolutions per minute as previously discussed. There are three sets of induction coils on the generator, positioned at 120° to each other. This was Nikola Tesla's design. As the electromagnetic field generated by the rotating element of the generator sweeps through them, it 'induces' a voltage across each of the three induction coils, each coil producing a sine wave with a phase difference of 120° to the other two in what is called a 'delta coil' or 'star connection'. See Figure 8.5. This is distributed as three phases right up to the domestic supply substations. In any town, the supply to different houses on the same estate would also be on a different phase of the supply. This distributes the required energy demand more evenly.

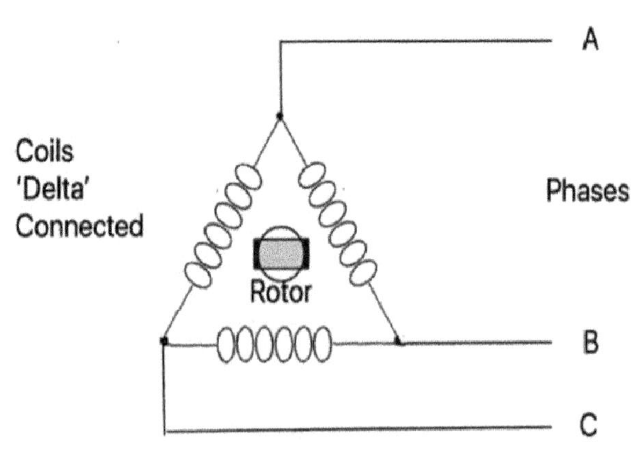

Figure 8.5 Delta coil formation.

In Figure 8.5, outputs A, B, and C would each provide one phase from across each coil, that is AB, BC, and CA, each 120° out of phase with the others. These could then be connected individually to single-phase transformers to step up or down to the required voltage or similarly be connected to delta transformers, where each coil would form a primary winding with an independent secondary coil stepping down to the required voltage. The other three-phase method is a star connection, where each coil is connected to a common reference point and the different phases are taken directly from the other end of each coil as shown in Figure 8.6.

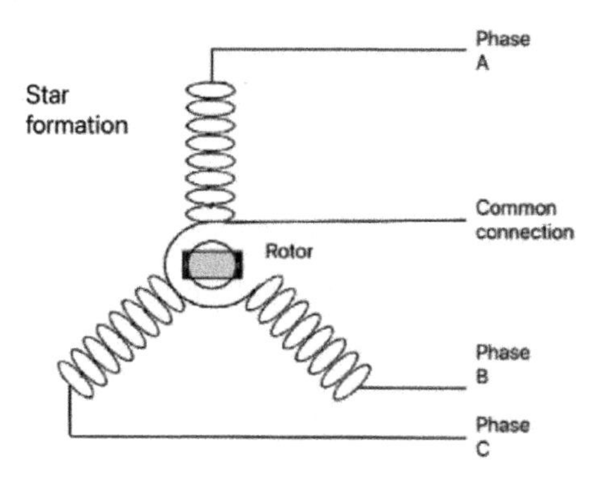

Figure 8.6 Star connection.

In summary, just one of the sine wave's outputs at the end user point is 240 volts, although as mentioned the actual peak to which it rises both positively and negatively is 339.36 volts. However, some industrial processes make use of 415 volts RMS. This is the difference in voltage between any two of the phases when reduced to the single-phase voltage of 240 volts RMS.

A dynamic magnetic field causes an electric charge within that field in accordance with Faraday's law. This states that an electric charge will occur in a time-varying magnetic field but not in a static one. A transformer makes use of this phenomenon by feeding a coil with the oscillating current to purposely generate a strong magnetic field that then induces a charge in another surrounding coil. The ratio of the number of turns in each coil steps up or down the induced charges as required. The amount of current that can be used in the second coil is the inverse of the voltage change. A higher secondary voltage delivers a proportionately lower current and vice versa. It is this principle that provides the very high-transmission voltages that can be further stepped up, and then stepped down, at point of delivery as the local domestic voltage as previously discussed.

In terms of the high voltages transmitted around the country at 50 Hz, there is a side effect, this being an oscillating magnetic field that is detectable around the transmission lines. The national grid network acts as a very large antenna for the 50 Hz fed into it. As discussed in previous chapters, the electric charge of several hundred thousand volts produces a magnetic field at right angles to it, and this oscillates proportionally in strength to the AC current.

CAN FIELDS EFFECT HEALTH?

Concerns have arisen that these fields are detrimental to the health of people living close to energised high-voltage power lines. These concerns are about strong magnetic fields being produced along the high-voltage wires and oscillating along the length of them. Many new housing developments are built very close to the national grid high-transmission lines, and there was a suggestion that incidence of childhood leukaemia was higher in these areas than what would be expected in the general population. It is true that these transmission lines produce oscillating magnetic fields that are detectable at some distance from them. The possibility of such levels of induced voltage adversely affecting human tissue or blood cells is virtually zero.

Anomalies that suggest there are clusters of childhood leukaemia cases within housing developments close to national grid power cables should be viewed in concert with the fact that new-build housing areas are more likely to have a far greater number of children. This is because younger families relocate to, and live in, these newer developments. When the reported cases are averaged out nationally, the number of cases probably remains within the norm for the whole population. It should also be noted that 50 Hz is not an ionising frequency capable of DNA interference that could possibly cause cancer of any sort.

Ripples in the Ether II

During my medical postgraduate research studies, I had to drive several times per week to the university and the teaching hospital nearby. On one straight stretch of the motorway, high-voltage wires can be seen to cross over at right angles to it from just over one mile on approach to them. This is not unusual and is the same for hundreds of crossings over roads all over the country. However, on one occasion when my car radio was set to a poorly tuned AM station, a low-frequency interference whistle started to develop. I noticed this from about one mile's distance from the wires, with further increases in volume as I got closer to them, rising to a maximum as I passed under them. This noise diminished as I then drove away. This shows that there is an effect detectable at a distance by the alternating field produced along the high-voltage cables. However, it should be noted that radio receivers amplify the signals induced into their aerials up to several million times. The detected signals are in microvolts. It is likely that the interfering induced voltages from the power cables are of the same order of being just microvolts even when closer to the cables.

Safety aspects of electricity distribution where live wires are to be found requires that precautions have to be taken at voltages of around 35 volts AC and 50 volts DC, as above these the current caused to flow through the body starts to interfere with bodily processes, including muscular contraction, and possibly including cardiac interference. The above-mentioned voltages again reflect the peak voltage difference between AC and DC supplies, as discussed.

The whole basis of these chapters on nonphotonic electromagnetic energy has been to emphasize that there is only an induction effect when this energy has frequency and is not static. The same principle applies whether at the high-frequency short-wavelength end of the spectrum or right down to those wavelengths discussed in power distribution. This serves to show that these oscillations all have a profound effect on our modern lifestyles and are mostly taken for granted.

Chapter 9

FREQUENCIES FOUND WITHIN THE BODY

In discussing the body, we invariably mean the human body, but similar discussions could be applied to most mammalian animals and other forms of life that have similar neurological systems. Although plants are living entities, they do not appear to have nerves in the same sense as animals do, so plants are not included in the following discussions. Although there must be some rudimentary signalling going on, especially within carnivorous plants such as the Venus flytrap, for this chapter the primary focus is on humans. The interaction with various frequencies is obviously keenly felt by humans, but these frequencies also affect the internal frequency generators within the body. The most obvious external one of these is the earth's rotation on its axis whilst further orbiting the sun. Within the body there are many frequencies generated autonomously from the very low sub-unity ones found at the lower-end frequency, generated by the brain, to those now believed to be around 1 MHz, also found within the brain as part of interactive consciousness throughout the brain. This chapter may only skim the surface of internal frequencies but attempts to give the idea of the vast range required to allow life to function.

CELLULAR MEMBRANES

The idea that there could be electricity within the body in any form resembling AC or DC types had long been dismissed by the majority of those within the medical profession. However, there has also long been an acceptance of the electrochemical nature as to how the body, and particularly the nervous system, operates. The cellular nature of tissue requires that an electric charge be maintained across the membrane of each and every one of many billions of cells making up human and other animal bodies. The membranes are micrometres thick, being composed of phospholipids arranged in two layers such that the tails of the molecules are in between the layers. This arrangement forms the majority of the cellular nerve membranes, acts to keep the organelles within cells, and adds an external structure to both cells and nerves whilst allowing the passage of water freely in and out. This leads to the term *aquaphilic* being applied to the membrane. See Figure 9.1.

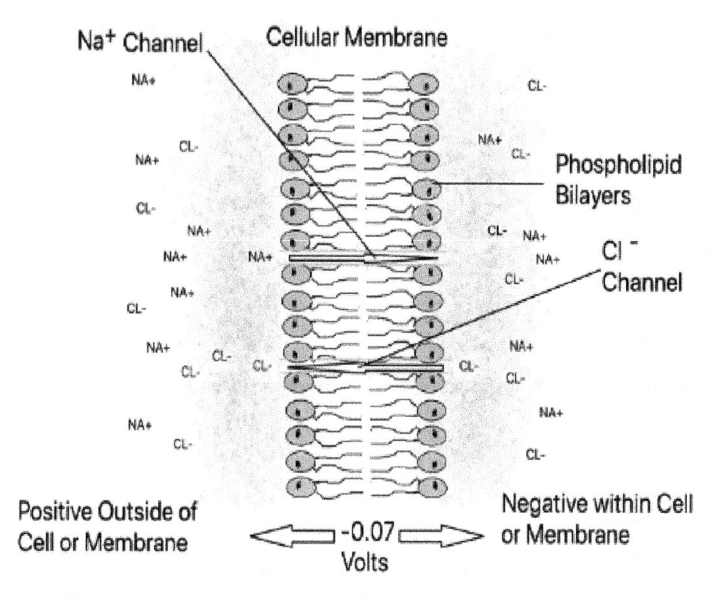

Figure 9.1 Phospholipid bilayer. These form membranes around cells and along all axons.

There is a measurable voltage across the membrane, maintained and established by an imbalance of ionic molecules called negative anions and positive cations. Slightly more anions are within the cell and migrate to the cell wall, attracted by the number of cations in the surrounding interstitial (intercellular) fluid. This voltage is generally in the order of −70 millivolts, which is less than one tenth of a volt. However, this small voltage represents a voltage gradient of several million volts per metre if all the membrane layers were stacked up to a depth of one metre and all the 70 millivolts were added together. The 70-millivolt charge is necessary to attract anionic nutrients into the cell and expel cationic waste ions from the cell through selective channels across the membrane. The transit time is in the order of one millionth of a second to cross the membrane, and collectively this would represent a very slight membrane voltage pulse of around one microsecond in duration as the exchange takes place. This one-microsecond pulse may be significant when looking into the conscious brain frequencies discussed later in this chapter. If flowing in and out of each cell at a constant rate, depending upon the requirements of the cell, the overall rate of pulses determined by the rate of exchange would constitute a frequency. A small pulsing electromagnetic signature per cell could probably be found collectively in the form of a randomised noise from all the billions of cells active at any one time. However, the cellular voltage discussed above represents a fixed voltage that could be described as DC (direct current) and that is similar to the voltage of a battery, attracting the cationic nutrients in and anionic waste nutrients out of the cell membrane.

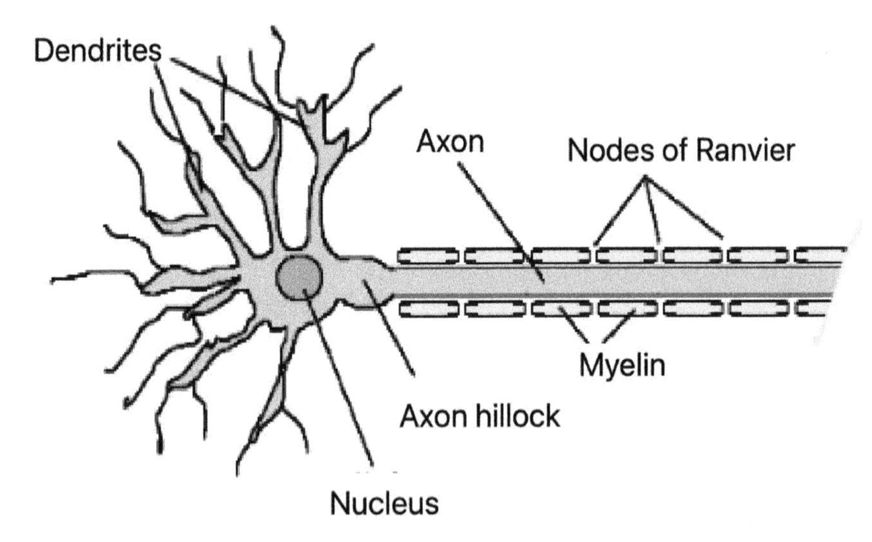

Figure 9.2 Neuron and its axon.

The cellular membrane feed of nutrients is the same for all cells, whether found in soft tissue structures such as muscles and organs within the body or in neurons within the brain. A neuron, however, has extra appendages in the form of inputs called dendrites. These resemble the branches of a tree and lead to a single output called an axon. These structures are also made up of bilayers of phospholipids with selective channels along them but are elongated. Called nerve fibres or axons, they are found throughout the body. See Figure 9.2.

Whilst the axons and dendrites are maintained by the parent neuron through a system of microtubules found within the centre of the fibres, the membranes surrounding the fibres utilise the membrane potential to attract sodium through selective channels into the axon in a way that distributes commands around the body called 'action potentials' (APs) introduced in Chapter 1.

The sodium ripples along the axon, entering through selective channels called 'nodes of Ranvier' and is then quickly expelled back into the interstitial fluid once the action potential has passed. The whole process can be compared to a Mexican wave. These action potentials have a frequency of repetition from very few to many thousands per second but are particular to the different types of nerve fibre. They are required to stimulate muscle contraction, and the frequencies depend upon the load being applied to the muscle. Where injury inhibits natural stimulation of muscles, artificial electrical stimulation is used by medical professionals to cause muscle contraction. This contraction requires a chain of electric charges applied along a target muscle at frequencies that mimic the natural rate of action potentials initiated by the brain. Voltages in excess of 100 volts are used in processes that stimulate muscles. For safety, they are applied in isolation to the mains supply and directly along the target muscle. They are of short duration and usually applied in groups pulsed at up to 220Hz.

REDUCING PAIN TRANSMISSION.

Short-duration high-voltage pulses are used to reduce pain. The pain gate theory describes how pain can be blocked by a mixture of frequencies in the form of pulses by electrically stimulating class A acute nerve responses after an injury has occurred. For example, when touching a hot object, the initial acute pain response is quickly transmitted from the injury site to the brain, eliciting a withdrawal action from the offending stimulus. Chronic pain arises a short while later as slower class C nerves are stimulated by prostaglandins that are released from the site of the injury and remain for quite some time after the initial injury. The pain gate system does not allow both chronic and acute pain at the same time as the two types of pain are transmitted along a shared ascending (afferent) neural pathway to the brain. It is the repetitive rate of pulses that is of importance to achieve maximum relief of the chronic pain. Stimulating the acute pain nerves at the same time that chronic nerve activity is present is achieved electrically by devices referred to as TENS (transcutaneous electrical nerve stimulation) machines. These effectively cause 'gating' of the chronic pain by electrically stimulating the acute pain pathways from a recovering injury, thereby blocking onward transmission to the brain since both pain signals cannot be transmitted at the same time.

As discussed above, chronic pain is just that, chronic, but it can be present in various levels according to stimulation by neural transmitters present in the areas of injury. These can range from a tick-over of one pulse (action potential) per second to highs of up to thousands per second. The brain interprets the rate of stimulus as the level of injury and with this identifies it as pain level. Eventually this can be saturated, and any further increase by the speed at which the nerves communicate with the brain is limited. Other uses of frequencies within the nervous system arise out of the stimulus of various nociceptors sending signals to the brain to reflect their state of stimulus, this whether high or just ticking over.

ACTION POTENTIALS

I have already mentioned that class A nerves carry acute pain signals at high speed, whereas the smaller class C nerves transmit at much lower speed. The repetition rate of action potentials is limited until the preceding one has completed its full cycle, called the absolute refractory time. See Figure 9.3.

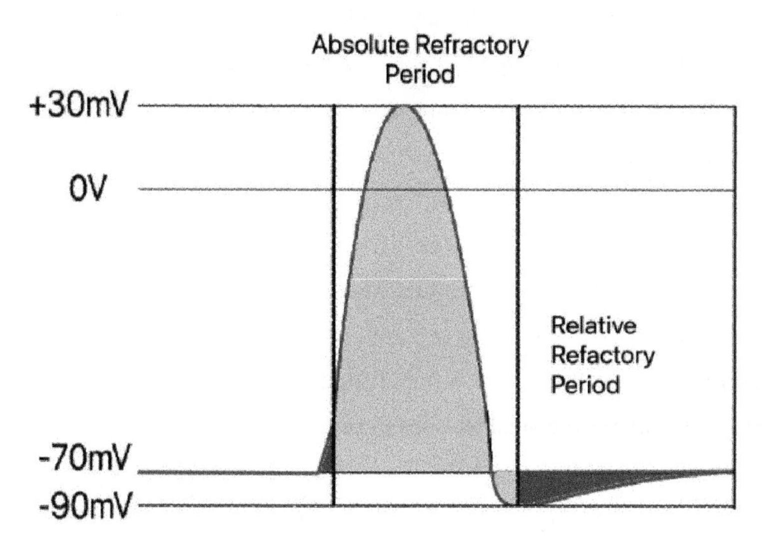

Figure 9.3 Absolute refractory time.

Figure 9.3 shows that for various nerves the rate of action potentials generated from the nociceptor (specialised sensory neuron) is limited by a specific time before another one can be initiated. This inhibits overstimulation and frequency of those being generated too quickly at the start of the chain. The frequency of action potentials (APs) generated by the brain can vary depending upon the requirement of the body. Muscles will simply twitch if the frequency of stimulative APs is too low. To sustain a contraction requires a higher frequency rate to cause the muscle cells (sarcomeres) to slide together. Even more APs are required if resistance to movement is encountered, such as when lifting a heavy object or sustaining tension to hold a weight firmly in place.

Ripples in the Ether II

Frequencies involved in the body cover a wide range and are required to build, control, and repair. Some of these frequencies come in sustained regular chains and others in short duration bursts. The following discussion will attempt to unravel these frequencies that are essential and required to sustain life itself.

Frequencies external to the human body range from ultra-high to very low in the sub-unity range, the latter taking millions of years to complete one cycle. However, although our senses are tuned to detect high frequencies such as light and heat, we have internal clocks and pulse generators triggered by external forces, producing both information and physiological reactions essential to body processes. These forces are sensed and detected by sending nerve impulses that are electrochemical in nature called action potentials (APs) that are transmitted back to the brain. These APs are caused and activated by specialist sensory nerve endings in tissue which are specifically stimulated by electromagnetic energy in the form of photons and by physical mechanical energy. Some of these were discussed in previous chapters.

IMPORTANCE OF FREQUENCIES IN THE BODY.

In Chapter 2 of this book, I attempted to calculate a frequency that I considered the lowest one possible, that is of the oscillating universe rate, but then discussed the incredibly high frequencies that emanate each with the same starting point, the Big Bang. Frequencies in the body start from the relatively low and synchronised, from diurnal clocks governing sleep and wakening activity, to brainwave patterns giving rise to our psychological profiles and the rate of nerve impulses interpreted by the brain to give sense, perspective, and reason to the world environment we find ourselves in.

From the very start of life in the womb, the embryo develops a beating heart. This is started and controlled by a neurological clock called the sinus venosus, better known as the sinoatrial node. The sinus venosus is a group of cells which form a frequency-controlled oscillator that can generate a continuous stream of action potentials that radiate over the developing heart muscle, which is composed of specialised muscle cells, causing them to contract. The term used is to 'depolarise' the heart's muscle, where the relaxation is then called 'polarisation'. See Figure 9.4, which is a cardiogram of my own heart, where from the start of the heartbeat cycle, it is divided into *p*, *q*, *r*, *s*, and *t*. Representing the various stages of the cycle, these can be used to assess cardiac function and stress and can change when the heart muscle is damaged during a heart attack, also called 'myocardial infarction'.

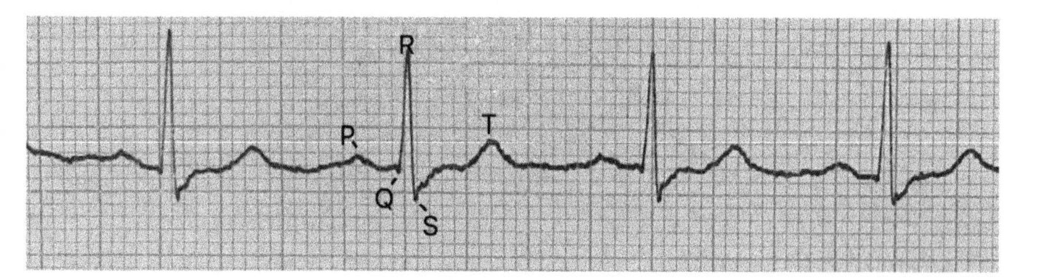

Figure 9.4 Cardiogram example.

These muscle cells are genetically programmed to form into four chambers that sequentially pump blood to carry essential oxygenated blood around the developing embryo and to transport carbon dioxide back to the lungs. The frequency of these embryonic heartbeats is between 120 and 160 beats per minute. These beats continue through life and slow down to a resting rate in a healthy adult of around 60 to 80 beats per minute. It may be that at some point in the past, where time values were being developed, the heartbeat was taken and measured at the pulse to provide a timing mechanism used for various processes that required some precision and gave rise to the period now called the second. The rate per second or frequency is named a hertz after Heinrich Hertz, whose name coincidently translates as 'heart' in German. It is also coincidental that 60 beats per minute is the average resting heart rate of a healthy human male.

I could further speculate that collectively counting the number of seconds in a day, which is around 86 400, coupled with timings related to the earth's rotation and mathematical divisions that are also in multiples of 60, this number is highly significant. Perhaps the degrees and measurements of the earth's rotation came from ancient timing by counting average male heartbeats for a complete rotation of the earth in a precisely measured length of a day, with 60 beats equalling 1 minute, then with 60 minutes equalling one hour. The figure 60 became the 'second' division of the minute after an hour was divided by 60 to give a time value to a minute. The term *second* was then adopted universally as the lowest standard reference integer unit time frame. All frequencies in this book are referenced to it.

The divisions of the day into hours initially came from ancient Egypt, where the night was divided into twelve-time segments and the year set at 365 days long. This became incorporated into the Julian calendar by Julius Caesar. In mediaeval times, the time between sunrise and sunset was also divided into twelve time periods, but this meant that in winter each of these period or hours was shorter than in summer. In the fourteenth century it was settled that the mean of these twelve individual time periods should be set as a standard hour regardless of the time of year. This gave rise to the 24-hour day as we know it, set in equal divisions through day and night. Sixty then divides nicely as the first *minute* division of an hour and again sixty as the *second* division by 60 of the minutes.

Heartbeats are but one of many frequencies within the human body, but these are not fixed as the rate varies dependent upon the body's demands for oxygen and for other essential nutrients carried by blood. Since this book is dedicated to frequencies, it would be a distraction from its main objective to discuss all biological processes that depend upon the beats of the heart. However, the heart and the blood it pumps allow other functions to occur that depend upon a series of oscillations to both control and regulate essential processes.

NEURONS AND THE BRAIN

Starting with the brain, it is composed of many billions of pulse generators called neurons. Initial estimates put the total at around one hundred billion, but that figure has been revised to be somewhat lower. These each are a sort of conditional pulse generator in that they generate an action potential because of weighted input conditions. Within the brain they are mostly fed by multiple inputs of 'enable' or 'not enable' from the output of other neurons. They feed to branches of each neuron's dendrites with many hundreds or thousands of interconnecting arms. The points at which they connect are called synapses. The receiving neuron then weighs the 'enabled' from the 'not enabled' at each synapse and, depending on the balance of inputs, produces an output action potential that is transmitted away from the neuron and along its axon. See Figure 9.2.

Given that the number of neurons in the human brain is between 86 billion and 100 billion (or less, as mentioned above), we know that these are not the only neurons in the human body. All sensory receptors are, in effect, neurons, these being stimulated by interaction with external influences and neurotransmitter chemicals, and each individually generating action potentials as discussed above. In most cases these potentials are then sent directly to the brain via the spinal cord.

Electronic computers have a central clock that synchronises all processor functions of inputs and outputs of the machine since the transfer of digital data requires absolute synchronisation. The data fed to a computer is in serial form but then buffered into registers that can contain up to 128 bits to 256 bits (binary digits) or more, each providing billions of individual patterns of 1s and 0s. Many of these patterns or 'codes' are then identified and given various predetermined functions. These are decoded electronically, causing different mathematical functions and responses when the especially designated patterns of 1s and 0s are recognised by pattern-identifying circuits. All this takes place in what is called the central processing unit (CPU), with each step taking its timing from that central clock.

In the human brain there is no known centralised clocking system since both input to, and output from, the brain through the central nervous system is controlled up to this point by feedback systems to muscles and from nerve endings forming sensory receptors. These sensory systems are necessary to allow humans and animals to safely exist in an otherwise hostile environment. Every time a neuron generates an action potential, it gives off an electrochemical pulse that can be detected by sensory electrodes placed at the skin's surface. How it achieves this arises from an electric charge at the beginning of the axon. This originates from the axon hillock, which resembles a swelling at the start of the nerve fibre (axon). As previously explained, when this charge reaches a threshold, a cascade of sodium enters the axon through specific channels and ripples like a Mexican wave along the axon and away from the neuron. Also felt as a charge of up to 1 mV radiated around, this can be electronically detected at the skin's surface.

The neuron is not just a decision-making unit but also a chemical producer of acetylcholine (ACh) and, in some neurons, gamma-aminobutyric acid (GABA), along with other neurotransmitters. ACh is essential for the transmission of action potentials across synapses. It is transported in microtubules within the axons in small packages, called vesicles, through a system called 'slow axonal transport', to arrive at the synapse. The action potential arriving at the synapse then opens specific channels to release the contents of some of these vesicles into the synaptic cleft. ACh then crosses the cleft and is attracted to ACh receptors on a neuron at the other side of the cleft. See Figure 9.5. This is called the 'postsynaptic neuron', which when sufficiently stimulated, and depending upon the type of originating neuron, generates another action potential that is then either transmitted along its own axon to another neuronal synapse or goes on to fulfil some other function of the body.

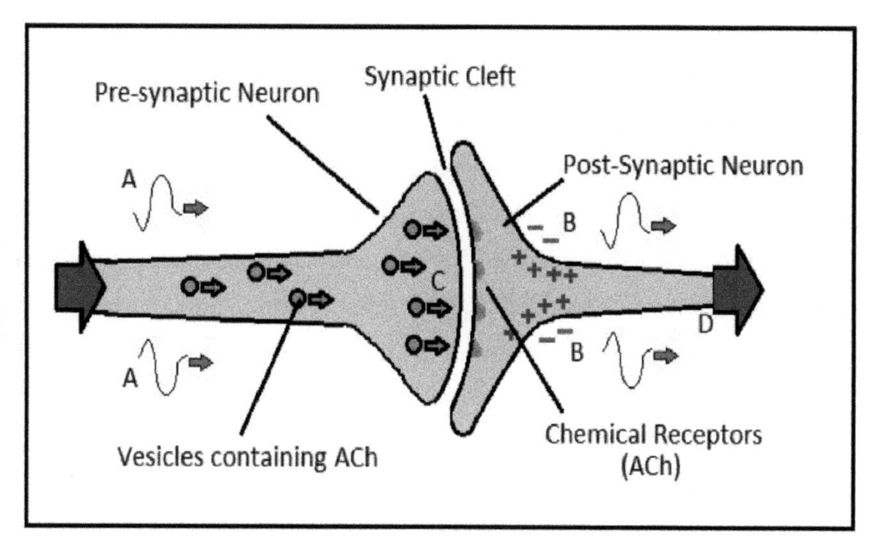

A = Action potential from neuron. B = New post synaptic action potential

Figure 9.5 Chemical synapse. The new AP is formed in the postsynaptic neuron.

GABA is an inhibiter used to stop onward transmission of action potentials. A typical use of this neurotransmitter is in the spine, where GABA plays a major part in the selective transmission of pain signals and is produced by specific types of neurons called 'interneurons'. It is possible to view neurons as selective binary switches, but the addition of GABA-producing interneurons provides for a tertiary option of null output for specific neuronal circuits, effectively closing the neuronal responses to all input APs.

The axon hillock at the start of the axon is the point where the balance of 'enable' or 'not enable' action potentials arrive at the neuron via its dendrites (input axons). These potentials are electrochemically transferred to the neuron by synapses that build up the charge until its level reaches a threshold potential at the axon hillock. This occurs if there are enough enables to outweigh the not-enables. In the case of sensory nerve endings, these are stimulated by neurotransmitters or by other stimuli, including mechanical and thermal inputs. These, like all neurons, initiate action potentials in the same way as throughout the brain, and exist in such numbers that to calculate a specific frequency is difficult. However, as each action creates a detectable electrochemical spike, my estimate of spikes per second throughout the brain is in the order of 10^{15}, which is a thousand trillion. This, if given a clocked equivalent, is far higher than any electronic computer yet available or likely to be for a very long time. This does not mean that a frequency of a thousand trillion, 1×10^{15} (Hz), is a specific frequency. If such were the case, it would mean that each action in the brain was processed sequentially.

MICROTUBULES, ELECTRONS AND OTHER ACTIVITY AND THE BRAIN.

The brain is a multitasking organ with the many different parts processing inputs autonomously. The real mystery to me is how it all comes together. Where is the central part that collates sight, sound, and all the other senses working with memories and prior experiences to formulate decisions that become an individual independent thinking entity? There was a suggestion several years back that there is more going on than the currently understood processes of neuronal communication throughout the brain. This was at the microtubular level (within the axon) within dendritic formations and neuronal interconnecting axons. Stuart Hameroff (Hameroff and Watt 1982), a quantum biologist and chief anaesthesiologist at the University of Arizona, suggested that highly active subatomic microtubular activity occurs when an individual is awake and conscious but that this is different when unconscious. Recent research has confirmed that there is indeed quantum activity within neuronal formations in line with the previous paper written more than thirty years ago. This suggests that such activity is present and part of consciousness. This paradigm attracted much criticism at the time.

Ripples in the Ether II

Temperature increases within microtubules have been found to increase electron activity, with detectable frequencies of 1 megahertz (1 MHz). A research team led by Anirban Bandyopadhyay (Sahu and Bandyopadhyay 2013), a preeminent researcher in the quantum science of biology at the National Institute of Material Sciences in Tsukuba, Japan, suggested that these microtubular resonant oscillations interfere and interact with others throughout the brain. It is further suggested that these give rise to familiar brainwave patterns, discussed below. There are very large amounts of electrons transiting in and out of atoms throughout the body at any given time, termed 'free' electrons, but within the dendritic and axonal microtubules they are suggested to oscillate when in a conscious state. My own theories are that those oscillations are initiated within the very small confines of dendritic microtubules and may be the result of thermal agitation. The lateral dimensions may form electron resonance chambers. Increases in such temperature may be caused by various stimuli from circadian rhythm control centres such as the pineal gland and locus caeruleus.

This oscillation of electrons would generate correspondingly oscillating micromagnetic fields that may join up synchronously throughout the brain. This would cause inductions to generate additional electrochemical charges that are detectable and associated with conscious awareness. The 'soul' or self-aware thinking centre of a person may be a collective synchronisation felt throughout the brain, not necessarily by dendritic action potentials communicating between each specialist area, although this does occur, but through magnetic resonances detecting and synchronising with these oscillations within each major processing centre in the brain. It may not be generally detectable by scanners observing the various centres of the brain when active or detectable through MRI imaging, but by an overall collective resonance at the quantum level of oscillations at very high frequencies, permeating throughout the structure when conscious. This would link or encompass all the senses, feelings, and thoughts for the purpose of making decisions and taking actions for life. Understanding how these interactions make us self-aware is, and may always be, a mystery.

My own theory, based on the fact that microtubular electron activity is present in the conscious brain activity when coming out of an unconscious stage such as simply waking up, is that the sensory cortex can be stimulated from any source in any part of the body. That part of the sensory cortex initiates a 1 MHz oscillation that then electromagnetically induces further oscillatory activity in adjacent centres. I would also suggest that as the action potential arriving at the brain is basically a short-duration electrochemical pulse where sodium crosses the axonal membrane within one microsecond, it gives off a series of 1 MHz electromagnetic jolts that possibly initiate the onset of 1 MHz oscillations in the densely populated dendritic formations in the part of the sensory cortex targeted. These would quickly spread through electromagnetic induction until the brain is fully functioning and conscious. This suggests that each of the senses is capable of being a 'consciousness initiator'. This would also suggest to me that the central nervous system can be stimulated by the autonomic nervous system should the brain's unconscious monitoring of it be presented with a problem. My own conclusion on self-awareness is that there is no consciousness centre of the brain but that being conscious and being self-aware is a state of being of the whole brain and nervous system through electromagnetic oscillatory interaction.

Although this book has, up to this point, tried to put frequencies in descending order, particularly with those derived electromagnetically within the human body in discussing the interplay of bodily derived frequencies, many factors influence the rate or frequency. These depend on such things as location, for example muscle contractions, sensory responses from touch, heat, sharp instant pain, chronic pain, thought processes, arousal, mood, and excitement—to name just a few. Activity detected from within the brain helps identify some of the more prominent neurological processes associated with temperament depending upon stimulus at any given time. These activities can be sensed by placing electrodes over the scalp where wave patterns, in the order of millivolts or less, are present and detectable. However, although there appears to be no central clock synchronising and organising all functions of the brain and body, the electrodes detect noise of neuronal activity at the skin's surface that is not random but is in the form of waves of increasing and decreasing activity. These waves, although part of an increasing spectrum, are very low, starting at about 0.5 Hz and increasing to around 100 Hz, and are impressed on what appears to be the brain's background noise, which is possibly caused by the many other processes within the brain.

BRAIN WAVES AND FREQUENCIES

Before looking at the frequencies or waves of increase and decrease in that overall noise of the brain, it is important to understand that the terminology applied to them are referenced to each other and not to the range of frequencies generally discussed in this book, that is in terms of the human brain, infralow frequency starts at 0.5 Hz and ranges up to gamma waves, which are around 31 Hz to 100 Hz. The term *gamma waves* should not be confused with gamma radiation as discussed in earlier chapters.

The brain, being an electrochemical organ, consists of the many billions of neurons which are cells that intercommunicate with each other to control functions throughout the body. They achieve this through a system of cellular membrane electrochemical voltage potential changes called action potentials, which occurs through a system of interconnecting fibres, namely the axons, usually described as nerve fibres. Attaching electrodes to the brain and analysing the activity detected from cortical voltage changes shows that, depending upon activity and conditions, certain frequencies are more prominent than others depending on specific brain activity. These are generally classified as shown in Figure 9.6. Ned Herrmann (1997) suggests the following explanation of the divisions and wave pattern where these brainwave patterns are prevalent:

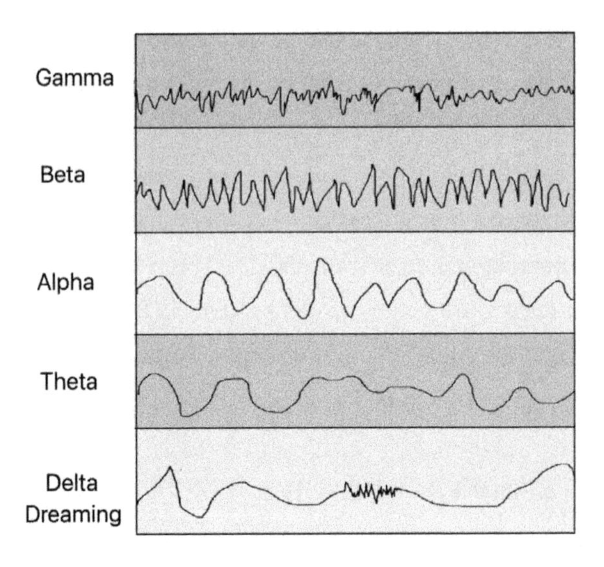

Figure 9.6 Brainwave frequencies (simulated).

- Gamma—31–100 Hz—indicates focus and expanded consciousness
- Beta—13–30 Hz—indicates normal working and concentration with frequencies up to 42 Hz
- Alpha—8–13 Hz—indicates a higher amplitude where the individual is thoughtful and relaxed
- Theta—4–7 Hz—indicates a drowsy state of contemplation or idealizing
- Delta—0.5–3 Hz—also indicates sleeping, but in a deep dreamless state

Delta (dreaming), 0.5–3.5 Hz, may indicate a sleeping state but one where the higher frequency is impressed on the low frequency, indicating that the individual is dreaming. Herrmann's view of delta waves is that dreaming is impressed on part of the delta waveform. Higher activity, or lower than average activity, within each of these frequencies may be symptomatic responses to problems and can be used to diagnose neurological problems.

FURTHER DISCUSSION ON PAIN

I am always amazed at the ability of the human brain to monitor every small point and detail of the human body. An Open University course I participated in entitled 'Brain, Biology, and Behaviour' suggested that every sensory nerve ending in the body sends a minimum of one action potential to the brain at a frequency of around one every second. This means that every one of those nerve endings is programmed to 'fire' an action potential without any other stimuli. These were described as maintenance signals. When stimulated by a neurotransmitter, the rate of action potentials increases. One morning I noticed that I had a minute wooden spell on the tip of my finger. The spell was so small that it could only be seen using a powerful magnifying glass. It was very painful even if the lightest pressure was applied over it. The size of it would suggest that only a single sensory nerve ending was being stimulated. That small spell was sufficient to increase the rate of action potentials from a single nerve ending to make it possible for the brain to recognise it and also enable me to pinpoint the exact location of the offending spell.

Having taught pain gate theory as part of the University of Nottingham's course in veterinary physiotherapy to postgraduate students, I was intrigued by the fact that there did not appear to be any chronic pain, as chronic should always follow acute pain responses. The dynamics of nerve transmission require that there be two types of nerves involved when a painful situation arises. These are loosely classified as type A and type C. Type A are the fast-transmission acute pain nerves, and type C are smaller and much slower chronic pain nerves.

It is my theory that whenever an injury occurs, there is a specific natural lag from the initial acute pain event to the time when the chronic pain kicks in. Typically, when a burn occurs, the withdrawal from the offending source is very quick. Only after a while does the chronic pain follow. It may be the case that both the chronic and acute pain receptors are triggered simultaneously, thereby blocking pain signals to the brain. Once the acute receptor saturates, this then allows the slower chronic signal to get through. In my case of the small spell stimulating a type A fast-transmission nerve, there was no damage to surrounding cellular structures, and because of this, no neural transmitters were released to cause a lower frequency and transmission rate of stimulus from type C free nerve-ending receptors that transmit chronic nerve responses. This would seem to conclusively prove several statements about the release of pain neural transmitters and also the pain gate theory. These pain gates are to be found in the dorsal aspect of the spinal cord. They appear to work in a similar way to computer logic gates and make use of GABA as previously discussed. See Figure 9.7.

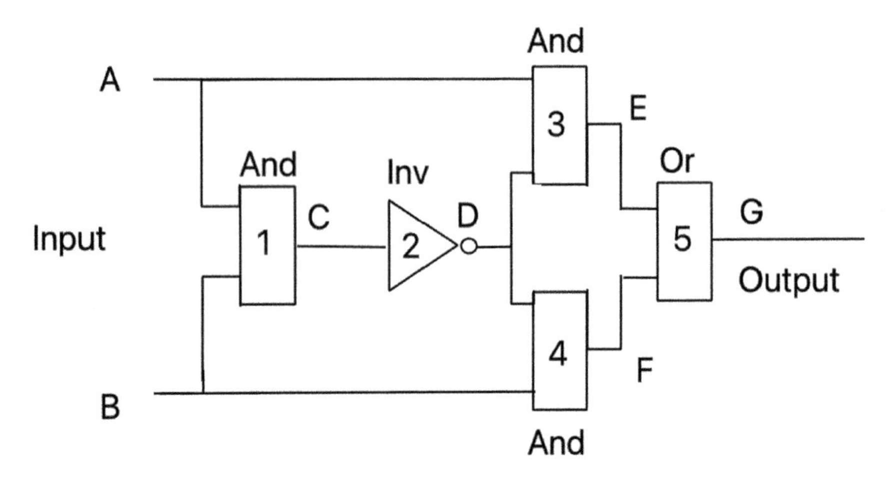

Figure 9.7 Logic model of a pain gate.

Figure 9.7 theoretically models a pain-gating system where inputs A and B represent acute and chronic nerve signals respectively. Logic gate 1 is an 'And' gate that requires an input on A and also an input on B to provide an output C. If there is only one input on either A or B, then there will be no output on C. Inverter gate 2 only gives an output on D when there is no output from C. Logic 'And' gates 3 and 4 are fed directly from A and B. Under no pain conditions, since there are no inputs to either A or B, the outputs from 'And' gates 3 and 4 will each have no outputs. 'Or' gate 5 allows an output when either E or F provides an input. If both A and B provide inputs at the same time, then logic C (representing the interneuron) prevents the inverter from giving an output on D. Only when A and B are individually active will the circuit allow the passage through to the circuit output G. This represents onward transmission to the brain. Inverter output D models the GABA effect.

Nerve signals sent from the brain to contract muscles are mostly controlled by conscious decision-making, or controlled unconsciously as a programmed withdrawal response to remove a part of the body further threatened by an external physical insult. Sending just one action potential to a muscle may have very little response, resulting in a very minor 'twitch'. It is only when the frequency of the rate of action potentials is increased that the muscle will start to contract.

The mechanisms of muscular contraction are well-documented and well-understood, but it is essentially a series of events that results in the contraction that is initiated by the action potential. Action potentials for muscle contraction are branched to simultaneously stimulate the whole muscle. This requires that a series of action potentials are sent serially to the muscle to cause a contraction. The strength of the contraction for lifting a load requires more action potentials during the dynamic stage but a still continuous, albeit reduced, chain to sustain the contraction if the load is to be held stationary under gravitational attraction. After a short period of time, the effect is saturated and the strength of the sustained contraction weakens. The rate or frequency of these pulses is not fixed but is load dependent, where the feedback mechanism to the brain initiates the correct rate to complete the action.

I studied in a postgraduate department where a project was ongoing to stimulate muscles in paraplegic patients. The idea was to build up the muscle mass and function such that the patient would be able to walk upright in a device called a reciprocating gate arthrosis (RGO). When the condition for muscle strength had been achieved, the patient, by using hand-operated triggers, would apply an electrical stimulus initiating a series of pulses to cause specific leg muscles to contract in such a way to support walking, thus making it possible. The muscle groups were stimulated by high-voltage pulses in a short series of bursts of frequencies around 25 per second. This rate of stimulation is probably in line with the rate of action potentials required for natural muscle contraction. The high-voltage pulses through muscle are isolated and therefore do not affect any other muscles other than the targeted ones.

Calcium is the key to trigger the chain of events that causes muscles to contract. Action potentials cause the release of calcium into the muscle from storage structures found throughout the muscle. The voltage changes cause special calcium channels in this structure, called the sarcoplasmic reticulum, to open. Pulsing the voltage from artificial sources also has the effect of opening these channels and causes the muscles to contract in the same way as brain-initiated action potentials do. In both cases, frequency is at the centre of the action and an essential component.

Not all voltages in the body are electrochemical. An American orthopaedic surgeon, Robert Becker (1998), found that nerves facilitated not only the transit of action potentials but also a DC-like current carried through the myelin surrounding the nerve fibres. He called these 'perineural currents', and they are manifest throughout the body where injuries or lesions occur and are now called 'currents of injury'. These were first identified and measured in the mid-nineteenth century by physician Du Boise Reymond (McCaig et al. 2005). How these charges arise is speculative, but I would suggest that the brain is the source by virtue of being formed in two halves. Each half has tens of billions of neurons aligned in a single orientation on the opposite side of the brain from the other. Since the brain is not evenly balanced, I would theorise that the aggregate total of the millions of action potentials being generated causes a charge to develop across the brain that is radiated through the neural systems throughout the body. This may be the source of perineural currents but may vary depending on how active the brain is.

This chapter could form the basis of a complete book on the current discussed above (using my theory) and may be dependent upon the amount of brain activity. However, keeping to the most obvious, the human body is programmed to the rotation of the earth. The most obvious indicator of this is visual with the detection of changes between night and day, along with effect of other stimuli that cause the release of specific hormones to control body functions maintaining synchronous harmony between day and night. This is called the circadian rhythm. It is believed that the pineal gland, found deep within the lower part of the brain, in prehistory played a part in regulatory functions by detecting daily changes in light. This gland is often referred to as the 'third eye'. This is more than speculative in that some animals, particularly reptiles, have a developed pineal gland closer the surface that is able to detect light and is constructed with common attributes found within eyes themselves. Pineal glands within the human brain are now incapable of detecting light and are a source of serotonin, which regulates sleep patterns in sync with not just changes in day and night but changes in season.

Chapter 10

THE SOUNDS OF LIFE

One of the frequency bands most obvious to humans is discerned through the ears. Anything that moves creates a wake in the fluid we call the atmosphere. It also causes waves much like those seen expanding two-dimensionally on the surface of water. Sensing these waves is a function of the ear's detectable range of frequencies. Humans are limited with regard to the audio frequency range, as will be discussed in this chapter, but electromagnetic waves or photons, discussed in previous chapters, have a commonality with sound waves in that the frequency can determine the energy. However, there are other factors, with the atmosphere being a very important one both in the transmission and absorption of sound energy, along with other mediums as will now be discussed.

At this point we now look at frequencies from a different perspective because they are no longer electrical or electromagnetic or forming any part of the electromagnetic spectrum. These frequencies are mechanical vibrations that now require an 'ether like' transmission medium.

An arts foundation course run by Open University and attended by my wife, Marjorie, posed the question during a summer school seminar 'If a tree falls in a wood and there is no one around, does it make a sound?' The answer to me would seem to be rather obvious. The falling tree is bound to cause mechanical vibrations with all sorts of frequencies transmitted as pressure waves through the atmosphere, radiating outward from the tree's impact. However, it requires that someone be present in the vicinity to hear these vibrations, which would be detected by the ears and fed as signals to the brain to be interpreted as sound. In other words, sound is a construct both of and within the brain and forms just one of the sources of frequencies discussed in the previous chapter. Sound vibrations are just a disturbance in the air and of no consequence in this context unless there is someone within earshot. I'm not sure that my interpretation of the question would have met with the approval of the tutors for that course, but it is the detection of disturbances in the local atmosphere caused by a transfer of energy from different sources that ends up as what we would describe as detectable 'sound waves'—but only if someone is around to hear them. Also, the question posed omits the possibility that other species are around that could also detect these sound waves and interpret them in the same way as humans.

SENSITIVITY OF HEARING

The ears are the main receptors for detecting sound energy in most, but not all, species, and the sensitivities are adapted for both hunting and survival. Confining the discussion to human hearing, we have specific and measurable limitations both in range and sensitivity. The purest form of sound is in the form of a sine wave, the frequency of which is called the pitch. With music, the basic pitches for the musical scales (see Table 3) are the same for all instruments. The scales are arranged around eight major notes from A to G, then starting again at upper A. This is a direct multiple frequency of the middle A, the frequency of which is 440 Hz. Table 3 shows individual frequencies of the main standard scales.

It is the harmonics adding to each of the frequencies (as listed in the table) and specific to the source that causes the 'timbre' or quality that is generated by each instrument. An A note played on a trumpet has the same pitch as one played on a piano, but the slight addition of added harmonics clearly differentiates one instrument from the other. Also, a mixture of frequencies in harmony with each other forms the chords that give warmth and quality to musical pieces. Other mixtures are discordant and, to most listeners, unpleasant to hear.

The accepted maximum frequency range of human hearing is from 20 000 Hz down to around 15–20 Hz. Above this range is termed 'ultrasonic', and below this is called 'infrasonic', or sometimes incorrectly called 'subsonic', equally meaning outside the natural human capability of detection. *Subsonic*

Musical note	Frequency
A	440 Hz
B flat	466 Hz
B	494 Hz
C	523 Hz
C sharp	554 Hz
D	587 Hz
D sharp	622 Hz
E	659 Hz
F	698 Hz
F sharp	740 Hz
G	784 Hz
A flat	831 Hz
A	880 Hz

Table 3 Musical scale (middle octave).

more applies to the velocity with which sound travels, meaning less than the speed of sound, whereas *supersonic* means faster than the speed of sound. With aircraft, the Mach number is given to the aircraft's speed with reference to the speed of sound. This changes with altitude as the air becomes less dense. The upper limit of human hearing diminishes with age, with only very young children able to perceive the higher-frequency end of the range. The speed of transmission radiating away from the sound source is variable. This is unlike the fixed speed of electromagnetic radiation in a vacuum. Also, sound waves diminish in intensity and are scattered at a rate dependent on the density of the intervening medium and distance from the source.

STRUCTURE AND ACTION OF THE EAR.

The ear, a masterpiece of design, is a structure designed to channel sound pressure waves on to the eardrum (tympanic membrane). These sound pressure waves vibrate the eardrum, which is directly connected to the first of three ossicles. See Figure 10.1.

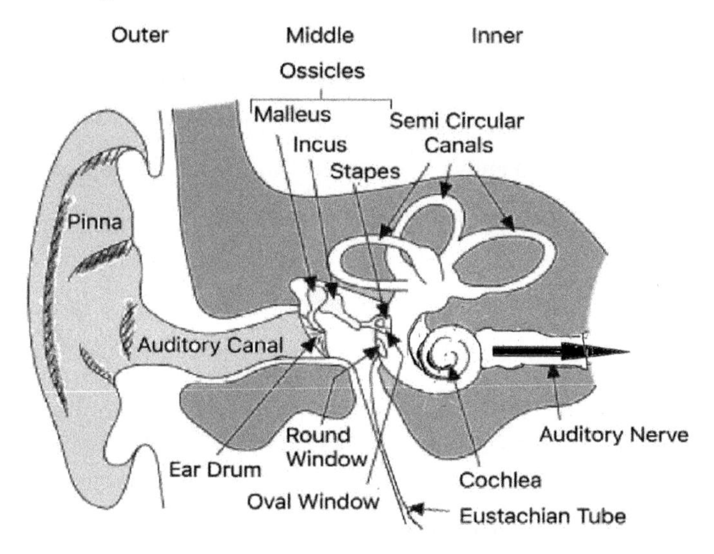

Figure 10.1 The human ear.

The ossicles are small bones called the malleus, incus, and stapes. They are commonly referred to as the hammer, anvil, and stirrup because of their respective shapes and are designed by nature to articulate with each other to give a mechanical advantage using a system of linking in such a way that physically amplifies the vibrations between the eardrum and the oval window. The oval window is the interface to a marvellous structure called the cochlea. This is a fluid-filled spiral formation that has resonant fibres throughout its structure, all tuned within its internal environment to a specific part of the human hearing range, with the frequency of maximum stimulation or resonance determined by the fibres' length.

The cochlea doesn't just detect one frequency at a time but can detect many hundreds. It is suggested that there are at least ten fibres capable of detecting frequencies which are between two adjacent notes on a piano. When the cochlea is stimulated, the motion changes potentials on the fibre roots and, in so doing, causes the generation of action potentials sent to the audio cortex of the brain. The brain processes these to determine the rate, intensity, and combinations of each stimulated frequency in order to identify both their source and their cause and, hence, identify meaning. This is processed in parallel with all the others to then allow the brain to sequentially decode and recognise patterns previously stored in the memory as a series. This identification process relies to a certain extent on sound patterns previously stored in the memory and associated with various sources and actions, and often locations and so forth, where comparisons can be made. I personally find that when driving my car through certain places, a particular song or tune will call up the memory of when I first heard it played at that location. This occurs even if the route is not a regularly used one with many years having elapsed before I returned to it. This same phenomenon also applies to conversations, possibly going back decades, but clearly remembered and associated with certain locations.

As a lover of music, particularly light classics, I am always amazed that the sound we hear is just one instant in time, much like the needle of a record player picks up that one part of the recorded groove to produce the flow of the sound. The sound that we hear is an example of the above-described process of sequential analysis by the brain. The wide stereo sound that we hear is again a construct of the brain, processing all those frequencies from the ever-changing patterns detected in the cochlea from each of those instants.

Stereo arises because the two ears pick up the same sound but at slightly different time phases and intensities, then the brain compares them and can then determine the location or at least the direction of the sound source. This give a stereo-wide audible picture with music, especially when microphones recording an orchestra or other sounds are placed on either side of the sources or musicians. The recorded channels are played back either through speakers or headphones, one to each ear. The process must also take from the memory of a familiar piece of music, voice, or other sound the expected patterns that cause the brain to compare the memory with the new input. If there is a deviation in that input of sound from that anticipated by the brain, then it immediately becomes apparent to the listener, eliciting an appropriate response.

In my time in industry, I worked alongside a colleague whose job was to develop pattern recognition systems. I became familiar with the complexities of the processes for analysing speech pattern power distribution for each of a number of spoken words. The programme was very complicated and had limited success in that usually only one person speaking could be interpreted by the system. In the intervening decades since then, the processes have been very much improved with most voice inputs for specific words now able to be recognised, but still limited in number when put into phrases.

The brain's ability to process and instantaneously identify almost a whole language or several languages, if learnt, with different pitches and accents is incredible to the extreme, as is its ability to selectively discern one particular conversation taking place in a noisy crowded environment from amongst all others around. This is done by processes as yet unknown. The process could possibly be related to anticipated and expected responses that the brain can discern from amidst the cacophony of noises arising in a crowded room and helped by visual cues of lip-reading and facial expressions between the two individuals conversing.

SOUND LEVELS

Power levels of audio frequencies derive their range from the amount of energy produced, which is caused by air molecules vibrating against the eardrum, about 1×10^{-12} watts per square metre being the absolute minimum for levels heard. The maximum short duration from below a thunderclap or close to a jet engine exhaust when at full throttle equals about 10 watts per square metre, well over the 120 dB threshold of pain to be discussed here. The sensitivity of the human ear is scaled logarithmically starting at that minimum level. The decibel scale sets the minimum at 0 dBA (A = acoustic) and the maximum at a point where permanent damage to the ear can occur after a short exposure of more than 120 dBA. These two extremes are respectively termed 'the thresholds of hearing' and 'the threshold of pain', and the recognised power ratio between these two levels is 1 000 000 000 000 to 1. Such is the intensity range of human hearing.

The minimum sound level possible of 0 dBA may seem incredible, but I can assure the reader that hearing is stimulated at this level. Whilst working for a university as a research technician, I had an anechoic chamber in my charge. This is a structure designed to be internally soundproofed by lining the internal floor, walls, ceiling, and access door with specially designed sound-absorbing materials. To make the chamber even more efficient in blocking out external sound, the whole structure was located within a reinforced concrete bombproof storage shelter at an old wartime bomb storage dump. It was further aided by being covered with earth and grassed over.

I was given a chance to experience what should be near total silence by sitting in the centre of the chamber. The feeling I would describe was that of increased air pressure with no echo or sound reflections to identify spatial orientation, especially with eyes closed or when all the lights in the chamber were extinguished. As I slowly became accustomed to this environment, it soon became apparent that the room was in fact not totally silent. A specific hiss could be heard caused by air molecules vibrating and colliding with my eardrums, clearly audible in the absence of any other externally generated sound. This was followed by hearing my own heart pumping blood throughout my body. The experience was one I found fascinating, but I have known of some people who found it claustrophobic and had to exit after a very short time.

It is safe to state that any environment within the atmosphere at temperatures above absolute zero can never be absolutely silent because of thermal agitation of air molecules. At this level the sound, in the absence of all other sources, is discerned as a 'hiss', as just described. This is wide-ranging and random and is called 'white noise'. As with all white noise, it contains within it almost all the audible frequencies detectable by the human ear, all being heard at the same time.

Audible sounds are not benign and can be dangerous to anyone overexposed to high-intensity power levels above the threshold of pain. A selection of typical sound levels caused by various sources in close proximity to the sources is shown in Table 4.

dBA	Source
0	threshold of hearing, as discussed, but also the sound of a heartbeat heard without a stethoscope
10	pin or very small object falling on a hard floor
20	a soft breeze through leaves
30	a human whisper
40	trickling water
50	a freezer or fridge motor
60	normal speech intensity in a quiet room
70	running water from a tap or shower
80	mechanical alarm clock
100	motorcycle engine heard by the rider without ear protection
105	roar of a small crowd at a football match
110	pop group heard from within the audience
120	pop group heard when close up to speakers; thunderclap

Table 4 Sound sources and levels.

The list in Table 4 is not exhaustive, and sounds much higher than 120 dB are encountered in everyday experience. A gunshot can be in the region of 140 dB but is of a very short duration.

It is obvious that this overexposure can damage the ear and the structures within it, but that is not all the consequences. High-intensity sound above 120 dBA can also directly affect the body. The energy in terms of power transfer may seem incredibly low. If rated in watts per square metre, then 120 dBA = 4 W/m^2. The following list shows the accepted tolerance levels of sound before damage occurs:

- At 91 decibels, your ears can tolerate up to two hours of exposure.
- At 100 decibels, damage can occur with fifteen minutes of exposure.
- At 112 decibels, damage can occur with only one minute of exposure.

In Table 5, we find the World Health Organisation's maximum recommended noise dose exposure levels in dBA.

Noise levels dBA	Maximum daily exposure time
85	8 hours
88	4 hours
91	2 hours
94	1 hour
97	30 minutes
100	15 minutes
103	7.5 minutes
106	3.7 minutes
109	112 seconds
112	56 seconds
115	28 seconds
118	14 seconds
121	7 seconds
124	3 seconds
127	1 second
130–140	less than 1 second
140+	no exposure

Table 5 Safe sound exposure levels.
At 140 decibels, immediate nerve damage can occur.

Table 5 does not mention frequencies and is taken for sound levels covering the whole spectrum. However, as with all energetic frequencies, the depth of penetration of structures and human or animal bodies is frequency dependent. Low-frequency sounds can penetrate structures and are less absorbed than the higher ones. The effect of high-intensity sound levels is to start to raise temperature levels within the body. This in turn increases heart rates to such a point that blood pressure increases to a level where it begins to rupture blood vessels. Sustained exposure can ultimately lead to permanent damage and death.

Beyond the human hearing range there are some similarities with EM radiation in that the upper limits of sound can reach into the frequencies to be found in low electromagnetic frequency (LF) wavelength bands. The upper frequencies used in certain sound transmissions are well over 1 MHz but are not generally found in nature. They are found in ultrasound equipment used to scan and image unborn babies in the womb or other visceral anomalies and also as a treatment modality used in therapy. It may seem inconceivable that a mechanical vibration is possible at such high frequency rates, but it is achieved because of the properties of certain materials. This is due to the 'piezoelectric effect' referred to in Chapter 8.

QUARTZ AND 'MOONSTONES'

One of the most ideal materials exhibiting this piezo-electric effect phenomenon is quartz, as mentioned in a previous chapter. To recap, quartz is a crystalline structure that when compressed will momentarily generate a voltage across the two planes of the crystal displacing electrons and forming crystal bonds such that the electron-depleted areas become positively charged. As the pressure ceases or is removed, electrons are quickly attracted back to the 'holes' created in the structure and, at the same time, give off a photon as part of the displacement energy, momentarily forming a charge in the opposite polarity. This piezoelectric property of quartz can also be reversed so that if a voltage is applied across the two planes of a crystal, it will momentarily change shape but quickly return to normal shape even if the applied voltage remains. If the voltage is pulsed and applied to a small specifically cut piece of crystal with a higher natural resonance, and if the electronically generated pulses match this resonance or a harmonic of it, then the crystal will vibrate at this resonance frequency, producing ultrasound mechanical waves.

Rocks sold as 'moonstones' are made of quartz which, when struck, can give off a flash of light. This light is the photons arising because of the displacement of electrons caused by the physical shock of the strike. These charges can be in the order of thousands of volts and are made use of to produce a spark in some gas cooker igniters and cigarette lighters. Some transducers used in mechanical sensors also use quartz to detect movement at far lower pressures and voltages, as this phenomenon is not restricted to sharp mechanical stimulation to achieve the effect.

The use of quartz to generate both ultrasound and voltages is made use of and encountered in almost every quartz clock or watch mechanism. One second, or divisions of a second, which is used as the standard reference throughout the world is produced by quartz crystals found within the watch's electronic circuit. The seemingly incredible accuracy of even the most inexpensive quartz watch is because of the piezoelectric property of precisely fine-cut crystals in the shape of a minute tuning fork and integrated into an electronic oscillating circuit. The natural resonance of the crystal in virtually all these circuits is 32 768 Hz. This is 2^{15} and a binary value of $2 \times 2 \times 2 \ldots \ldots \ldots \times 2$ repeated 15 times. Integrated dividing circuits, known in electronics as 'flip-flops', repeatedly halve the 32 768 Hz crystal frequency to ultimately produce the required 1 Hz. This is then used to time a digital clock display or to electromagnetically produce movement fed to watch or clock hands.

ULTRASOUND AS A TREATMENT MODALITY

With regard to using sound transmissions above 1 MHz for ultrasound equipment, when used as a treatment modality, specially designed applicators allow the generated ultrasound energy to be applied to tissue as single pulses or as a stream of pulses that is dependent upon application types and needs. A coupling water-based gel is required to reduce the impedance mismatch between the applicator and the skin surface. This simply means that most of the ultrasound energy will be lost as heat, allowing very little vibrational energy to cross into tissue unless a coupling gel is used. For imaging, a single pulse is sent into tissue and some of the energy of that pulse reflects off internal tissue interfaces. These reflections are detected at the applicator head. The timing between initial pulse and returned pulses provides a measure of the depth of the reflective tissue and also the maximum rate at which the pulse can be sent again in a continuous chain.

Along with some clever electronics and displays, if the ultrasound head is designed to scan at different angles, then an image of the internal structures can be achieved, including the state of foetal development in the uterus. Since these imaging ultrasound pulses are sent in a chain with a time delay set to allow the reflected ultrasound signal to be returned, a condition called cavitation within tissue fluids will neither occur nor cause a dangerous increase in tissue temperature. The water-based gel, as previously mentioned, is required to match the impedance (resistance of the tissue) to the active head for accurate imaging.

With therapeutic ultrasound, a stream of pulses of up to 3 MHz is used in much the same way as imaging as discussed above. The main difference is that there is a thermal effect as the ultrasound energy is absorbed by the tissue. Some reflections will occur from tissue-type interfaces when using the higher-frequency stream of pulses (short-wave ultrasound), and these can be problematic if the head is held stationary. Cavitation occurs when the reflected waves meet the incoming ones and form a 'standing wave' that can increase in intensity. This, if allowed to continue, can cause tissue damage but can be overcome by two possible methods. The first is by keeping the head moving at different angles over the application target area during the time of application.

The second is to pulse the chain, allowing bursts of ultrasound energy, which will reduce the possibility of cavitation occurring.

Long-wave ultrasound makes use of very much lower frequencies of around 40 KHz where the wavelength is too long to allow any reflected wave to establish the conditions for cavitation effects.

NON-MEDICAL USES OF ULTRASOUND

Like electromagnetic transmitted energy, the frequency of the pulses carries proportionate energy, making the higher rates of ultrasound frequencies (short wave) more energetic than the lower-end ones (long wave). However, the lower frequency finds use because it has greater depths of penetration but is less absorbed and scattered. Imaging ultrasound finds uses other than medical ones in measuring distances within solid structures to find potential problems such as faults and cracks. The speed of transmission in all cases of usage is dependent upon the density of the intervening tissue or structure. This, unlike with electromagnetic transmissions, means that wavelengths for specific frequencies are different, dependent upon the transmission medium to which they are applied.

Sound produced in the atmosphere of all frequencies and sources also varies in wavelengths as atmospheric density changes. At the standard reference pressure of 1013 hectopascals, the speed of sound is around 330 metres per second. A standard observation in physics is that the energy is dispersed or scattered by the medium it encounters. This applies equally to photons and sound waves alike. It also generally applies to all frequencies as to how quickly they are absorbed or scattered.

INFRASOUND

Infrasound is that which is not detected by human hearing systems because it is below the sensing range of the human ear, but it is often felt by other sensory mechanisms. The very low bass frequencies from high-quality speakers called 'woofers' can be felt as pressure waves by placing the hand close to the speakers when bass-frequency music is being played through them. These low-frequency sounds are very penetrating. An example is the bass frequencies that can be heard from oncoming cars where there is loud music being played inside. The bass frequencies are heard well before the vehicle is close by, whilst the high-frequency music component is largely contained within the internal confines of the vehicle cockpit. Another quality of very low frequencies or subsonic sound is that they can travel longer distances, especially through dense mediums.

Animals are known to be able to detect infrasonic sound waves emitted from deep within the earth. Plate-tectonic activity causes movement of the plates as they slide over one another. Occasionally there is a build-up of pressure, slowing down the movement. When this is overcome and suddenly released, the vibration caused is sent out as an infrasonic shock wave, which may then be followed by a physical manifestation of shaking of the ground or an earthquake. It can be speculated that this occurs more at times of solar a eclipse when the combined gravitational effect of both the sun and the moon causes extra stresses within the earth with increasing effect.

As previously discussed in this chapter, sound frequencies within the audible range can be dangerous if the sound intensity exceeds 120 dB, the threshold of pain. Sustained noises above a very much lower threshold of 90 dB can start to impair hearing by causing cochlear damage. At 120 dB and above, permanent damage can be caused, along with other effects. These include temperature increases on the skin's surface, increased blood pressure, and internal organ failure resulting ultimately in death. Sustained high-frequency sound such as that from jet engines, speakers at a pop concert (if one is close to them), or close proximity to a gunshot can cause a ringing sensation in the ears after the immediate effects of the sound have passed. Apart from the potential damage to the cochlea, excessive exposure to high-intensity sound levels at high frequencies may cause a feedback system to be established and sustained within the neurons found in the audio cortex of the brain. This is speculated to cause the establishment of a permanent feedback loop whereby the high frequency is continuously processed and fed back on itself in such a manner that it is heard by the sufferer as a high-pitched whistle. Such a condition is known as tinnitus. As a sufferer I have measured my own whistling tinnitus and found it to be between 9 KHz and 10 KHz.

WHITE NOISE

The similarity between sound and visible light is that if all frequencies of light are emitted simultaneously, then this is interpreted as white light. If all frequencies of audible sound are similarly heard at the same time, then this is called 'white noise', as previously mentioned, and is heard as a loud hissing sound.

Here we have a link between random-energy electromagnetic frequencies emanating from the original Big Bang and some current audio systems. If a radio is detuned or operating between selected frequencies, then a hissing sound is often heard from the speakers as the electromagnetic noise from all untuned sources produces micro-voltages in the antenna that are then processed by the radio's detection circuits. Some of this noise is thermal and comes from certain components used in the radio's circuit design, but the rest is picked up from induction into the aerial from those other sources, including space. This noise is largely filtered out with modern radios, especially as the digital age takes over, but older analogue systems may still hiss and, in so doing, provide the noise of Creation as a proportion of noise heard originating from the start of time. Distinguishing cosmic electromagnetic noise from local random noise heard is extremely difficult, but radio telescopes 'see' this background noise throughout the universe. The universe is very noisy throughout as all areas investigated by these telescopes detect this random energy as separate from earth-based background electronic noise.

USE OF SOUND FOR SURVIVAL

All the natural frequencies of sound are essential for development of species and their survival. Some species such as bats have developed systems similar to radar to locate objects and insects in totally dark locations by emitting short-duration high-frequency sounds and listening, with enlarged ears, for the echoes, with the brain forming a sonic picture of the bat's surroundings. An analogy for this would be humans emitting a short burst of light and picking up the reflections to see all things around. This may seem far-fetched, but in historical times this idea was put forward as an explanation for how sight worked, the believe being that the eyes were the source of the light.

Another use of sound in the breeding and, hence, the survival process is seen in a species of tree frogs. The male Borneo tree hole frog, *Metaphrynella sundana*, climbs trees to look for cavities up to 30 feet above the ground to make its nest, but not just any cavity will do. The frog emits a call at different pitches until it finds one that sonically matches the resonant qualities of the cavity, effectively amplifying the call. The purpose of all this effort is to attract a mate. A strong loud call suggests to the female that it originates from a superior male, ideal to mate with. Zoologists at Malaysia's Sabah National Parks (Lardner and Lakim 2002) reported this in an issue of the *Journal of Science*.

SONAR

Modern equivalents of sound devices used other than for hearing are things such as sonar, which is used to locate underwater objects and fish shoals and to range water depth beneath the boat. There are also inexpensive short-distance measuring devices that use a very short but high frequency pulse that can be pointed at a wall or solid object to measure one's distance from it. This is done by timing how long it takes for the echo to return to the same initiating transducer, which then works as a sensor. The absolute accuracy depends upon the air pressure at the time of measurement, the standard being 1013 hectopascals.

Ripples in the Ether II

With sound transmission through water, water being about eight hundred times denser than air at standard pressure, it means that transmission is much faster, but water is also more absorbing. As with the example of car speaker systems, bass or very low-frequency sound will penetrate to a much greater extent than the higher-frequency sounds. Animals such as whales emit sounds in the sea at low frequencies that can travel and be detected at great distances away. Whales also make use of sound, not only for communication with each other, but also for location finding and navigation. The speed at which sound travels in water is also greater, increased to around 1450 metres per second (more than 3000 miles per hour). This is faster than a bullet fired from a high-powered, high-velocity rifle.

Sound may feel like the poor relation in the scheme of all things regarding frequencies, but for life on earth it is very much an essential source of energy transfer within the atmosphere. It allows us to communicate uniquely as individuals with our own patterns of speech, to sing, and to relax as we listen to sounds that we find pleasurable. It can be a portent of doom at the infrasound frequencies, indicating a possible earthquake before any other manifestation occurs, but also an aid where alarms of many forms are sounded to take action to preserve life.

Brought up by a totally deaf mother whose perception of the world was governed by sight and her other senses, I respected my mother for carrying on as almost normal. But how much she missed out on life! Those who cannot hear are deserving of total respect for carrying on and being able to live and work without access to this most important ability to hear within the sonic part of this frequency range.

Chapter 11

SUN, MOON, PLANETS, AND GRAVITY

The earth is just one of many celestial objects orbiting the sun, with orbital rates determined by those objects' distance away. Nearly all the planets appear to have slightly elliptical orbits. These are more obvious with the nearer planets, but for those at a distance, the orbits are more difficult to perceive because these orbits take many years to complete. In physics, the rate of oscillations, or in the case of planets, the orbits, does not depend upon the planet's size. This is partly because the planets effectively orbit in a vacuum, but they also have a constant accelerating force, the sun's gravity. This chapter will open this up with the wonder of the simpler physics to give an overall understanding and give the orbits a frequency value in line with others already discussed, which will be added to later.

The frequency of repetitions applicable to the earth, moon, and sun are the most obvious to perceive and affect all life on earth. These are related to several major influences, mainly the orbit of the earth around the sun, the orbit of the moon around the earth, and to a much lesser extent, the orbits of other planets. At the beginning of this book, the unit of hertz (Hz) was adopted as the standard reference related to the number of repetitions per second. A simple mathematical illustration was given to show the number of seconds between summer solstices, and this value converted to hertz.

EARTH AND MOON

The most obvious and powerful influences determining the earth's orbit are the sun and moon, along with the earth itself spinning on its axis.

If we take the earth's rotation as a starting point, it is suggested that the spin was initiated about four billion years ago by a collision between another smaller planet and the earth. A theory is that the earth's gravity attracted most of the debris from this collision and the energy from the glancing blow caused the remaining earth bulk to spin. Debris, not all captured by the earth, may have eventually formed the moon from what would have been an accretion layer circling the earth—again at a rate caused by the collision. The energy of this debris arising from the collision caused the forming moon to rotate, and under influence of the earth's gravitational attraction, fixed the moon into a captured orbit such that one rotation of the moon matches exactly the rate at which it orbits the earth, this being equivalent today to 27.32 earth days. Obviously, the moon has its own mass-dependent orbiting rate and matches its own rotation so that the same side is always seen to be facing the earth, except for a very small wobble allowing 59 per cent of the surface to be seen from earth. The attraction of the earth's gravitational pull gives rise to what is called 'a captive orbit' that is in itself the result of earth's gravitational attraction gradient changing the centre of gravity of the moon from its absolute centre to the side, being pulled by the earth, and also a slight change of shape.

The 'barycentre' is not a physical point but a dynamic one and is the centre of mass for two bodies orbiting each other. The barycentre of the earth–moon combination moves slightly towards the moon as it orbits the earth. This agitates and stirs up the molten core of the earth, causing high core temperatures. The two gravitational orientations make the moon act like a pendulum on a gravitational string, and therefore it always presents the same side to us. This is different from the earth's rotation on its own axis, causing the seasons as it orbits the sun.

Figure 11.1 shows the apogee (142 000 000 km) being the farthest point of the earth's orbit round the sun, and the perigee (136 000 000 km) being the closest point. It also illustrates how the average angle of the earth's rotation gives rise to the seasons.

Interestingly, living coral lays down an identifiable layer of calcium carbonate every single day, giving a record much like tree rings can record years. Fossilised coral originating from many millions of years ago gives a record of 410 layers per year, meaning that there were 410 days per year at that time of its life. This does not necessarily mean the year was longer than it is today but that the rotation of the earth on its axis was faster. Relative to today, the number of hours in a day would have been around $365.25 / 410 \times 24 = 21.38$. This would suggest that in the earlier stages of the earth's development, it must have rotated much faster after the original collision before life became established on the planet.

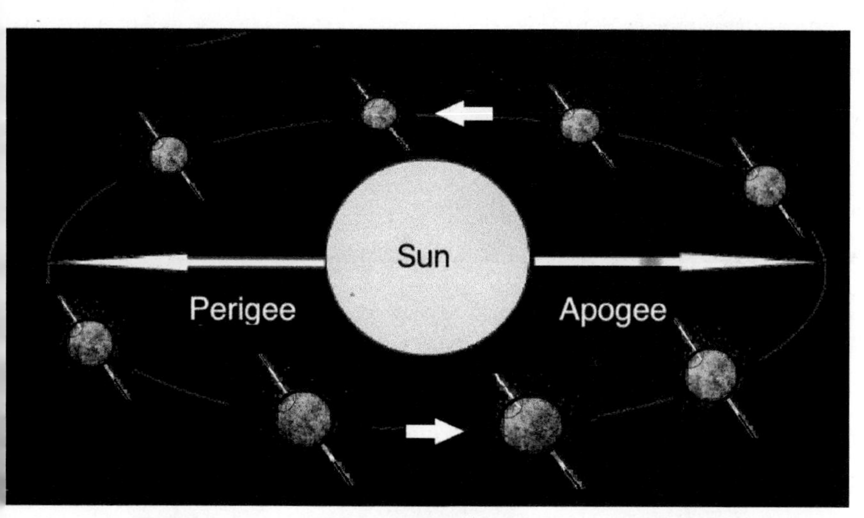

Figure 11.1 Oblique view (not to scale) of the earth orbiting the sun on its own axis due to the gyroscopic effect of the earth's rotation.

ECLIPSES

It is quite a coincidence that when the full moon is visible, it presents a visual disc diameter that is around the same dimension as the diameter of the sun. This is most obvious in solar eclipses. In total eclipses, the moon's disc completely covers the sun, giving rise to a period of total darkness in the sun's shadow. However, the moon's distance from the sun does vary slightly in its orbit of

27.32 days when at times the earth–moon combination is aligned to the sun. The moon's orbit around the earth is elliptic, so it varies in distance. When an eclipse occurs, the moon is at its apogee (greatest distance from the earth), and in a direct line with the sun in relation to the earth this results in an annular eclipse. See figures 11.2 and 11.3.

Figure 11.2 Almost a full solar eclipse photographed through the clouds. (My photo.)

Figure 11.3 An exaggerated annular eclipse.

This type of eclipse occurs when the moon's disc covers the majority of the sun's but leaves a ring (ring of fire) around the circumference.

The other form of eclipse, already mentioned in the introduction, is the lunar eclipse. This where the moon falls into the earth's shadow when aligned at the opposite side of the earth from the sun. It is often taken by some sects and religious groups as a portent of doom or some cataclysmic event particularly because the moon sometimes turns reddish in colour whilst remaining visible throughout the event. This is abject nonsense and can be simply explained by something that happens every day on the earth, namely the sunset. See figure 11.4. As the earth rotates, the sun sinks visibly lower down towards the horizon and appears red. The sun has not changed, but our perception of the light from it has because we see it through more of the atmosphere.

Figure 11.4 Spectacular sunset showing that the red-light part of the sun's spectrum is the last to show before setting. (My photo.)

Rayleigh scattering makes the blue sky visible above the clouds by way of the scattering of the blue photons, as discussed earlier in the book.

Earth's atmosphere scatters light travelling from the sun all around the planetary disc of the earth, perpendicular to the sun. The lower-frequency photons have to travel farther before illuminating the moon. It is these red photons which reach the lunar surface and are then reflected back to the observer on earth. This gives rise to the apparent red colour of the so-called 'blood moon'. See Figure 11.5. The variation of intensity would depend on how centrally located and how close the moon is to the earth at that time.

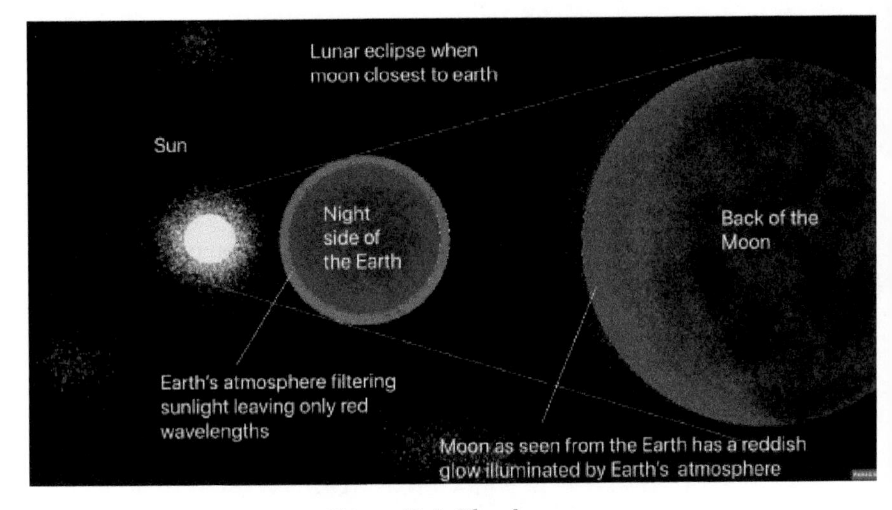

Figure 11.5 Blood moon.

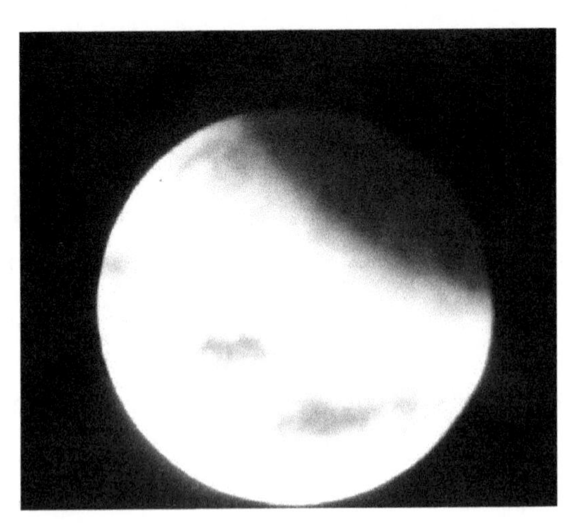

Figure 11.6 Lunar eclipse showing the earth's shadow across the moon's disc.

If a circle were to be drawn that followed the curve of the shadow, it would provide a good size comparison of the moon to the earth. See figures 11.6 and 11.7.

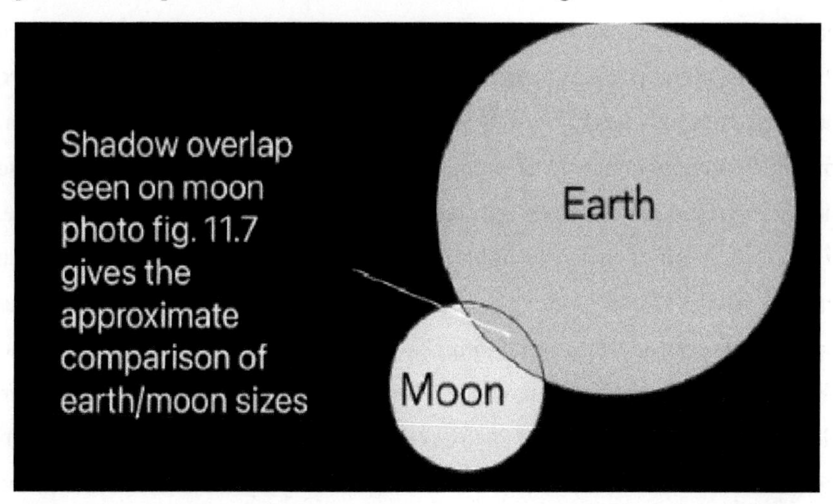

Figure 11.7 Relative sizes of the earth and the moon deduced from the earth's shadow.

The earth's orbit around the sun takes about 365.256 days. It doesn't follow a purely circular orbit but is also slightly elliptical. This means that on a midsummer's day, the earth is at its furthest point from the sun at a distance of 152 000 000 km. Conversely, at the northern hemisphere's winter solstice, the earth is at its closest point to the sun at distance of 146 000 000 km.

The question may be asked as to why, when the sun is closest to the earth, it is colder in the northern hemisphere. This is because the earth is tilted on its axis by 23.5°, meaning that the average energy received in the northern hemisphere is less per square metre than in the northern summer months, when it is closer to the sun. The maximum energy received from the sun when the angle is perpendicular to the sun is 1.336 kW per square metre, varying about 0.2 per cent in an eleven-year cycle due to sunspot activity. As the angle per square metre increases, less of that square metre is exposed to the sun, such that at 90° laterally to the sun's radiation there would be no sunlight absorbed.

The tropics are lines of latitude 23.5° north and south of the equator. The sun is less angled in the southern hemisphere and sits directly overhead the south's Tropic of Capricorn during the winter solstice. This means less of the northern hemisphere is exposed to sunlight per unit area as the angle increases by a total of 47°. Conversely, during the summer solstice, the sun is directly overhead the northern Tropic of Cancer, presenting a less acute angle to it above this latitude. Therefore, more of the sun's radiation is absorbed per unit area. Because the earth's orbit is slightly elliptical, it means that for the north there is about an extra week between the spring equinox and autumn equinox as compared to the southern hemisphere. This value of energy received per square metre from the sun is the 'solar constant' measured at the edge of the atmosphere at 90° perpendicular to the sun. It is reduced considerably by the time it reaches the earth's surface because of the atmospheric scattering and reflection back into space by clouds.

ORBITS

In our bid to put frequencies on these earth–sun orbits and earth–moon effects, we can calculate them as follows: the earth orbits the sun every 365.256 days. The number of seconds in a year was calculated earlier in the book as 31 558 118 seconds. This gives this earth an orbiting frequency of:

$$1 / 31\,558\,118 \text{ or } 0.317 \times 10^{-7} \text{ Hz}$$

For the moon's orbiting rate and, therefore the 'wobble' frequency that this causes to the earth, we have:

$$27.32 \times 24 \times 60 \times 60 \text{ Hz}$$

giving a frequency of:

$$1 / 2\,360\,443 \text{ or } 0.4236 \times 10^{-6} \text{ Hz}$$

This wobble causes the earth to be farther away from the sun when the moon is in full opposition and closer when approaching a lunar eclipse of the sun. This means that the energy arriving at the earth's disc constantly varies because of many orbital factors.

In Chapter 3, when discussing the orbit of the electron around the nucleus in a hydrogen atom, it was suggested that this caused a wobble in the neutron–electron combination. I also alluded to the fact in the introduction that in the earth–moon combination there is also a wobble, as discussed above. This causes the earth not only to follow an elliptical orbit around the sun but also to have a wobble frequency caused by the moon, affecting the earth as it orbits every 27.32 days.

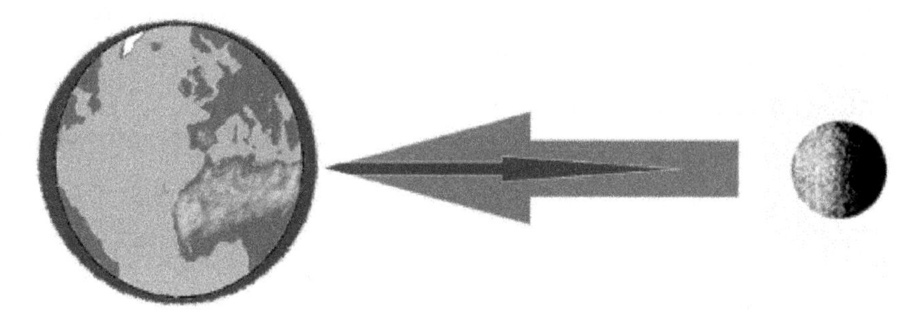

Figure 11.8 Earth–moon gravitational pulls

In Figure 11.8, the arrows indicate the approximate proportionate effects of gravity. The larger arrow is the earth's gravitational pull on the moon. The smaller dark arrow is the moon's pull on the earth that causes the tides and the wobble. The dark area surrounding the earth represents the seas bulging towards the moon, but these are greatly exaggerated. The bulge on the opposite side of the earth from the moon is caused by inertia on account of centrifugal forces arising from the earth–moon wobble, as discussed above. This gives us our twice daily tides.

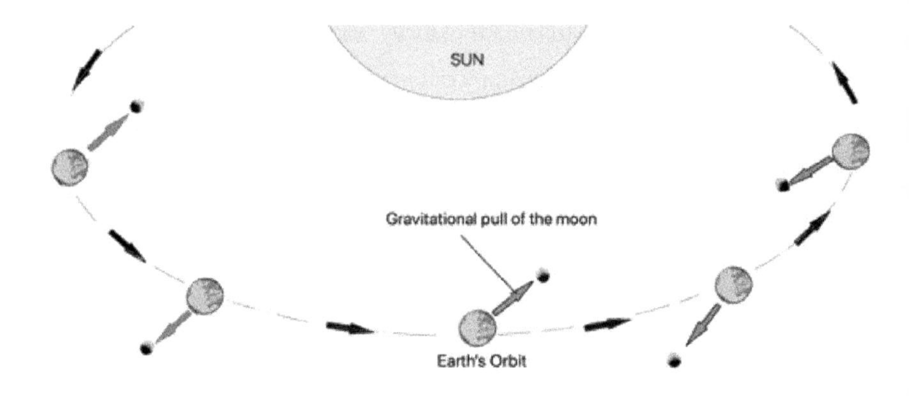

Figure 11.9 Earth–moon orbital wobble (not to scale).

Figure 11.9, though not to scale and greatly exaggerated in terms of size of the earth and moon compared to the sun, serves to show that the tidal wobble also causes a wobble in the joint orbit of the earth and moon around the sun. The centre of the wobble is within the earth's own radius at the barycentre, which offsets the gravitational centre of the earth towards the moon and likewise the gravitational centre of the moon towards the earth. With the moon, the gravitational centre remains offset and fixed, unlike the barycentre of the earth, which moves in a synchronous fashion with the moon's orbit.

The agitated molten core of the earth, caused by the gravitational effects, causes extremely strong electric currents to flow within it. This causes the earth to have an overall dipolar magnetic field with poles aligning at roughly 5.15° to the moon's orbit (see Figure 11.10). It could be suggested that the moon's orbit and gravitational attraction contributes also to the magnetic field that helps protect the earth from cosmic radiation and solar flares.

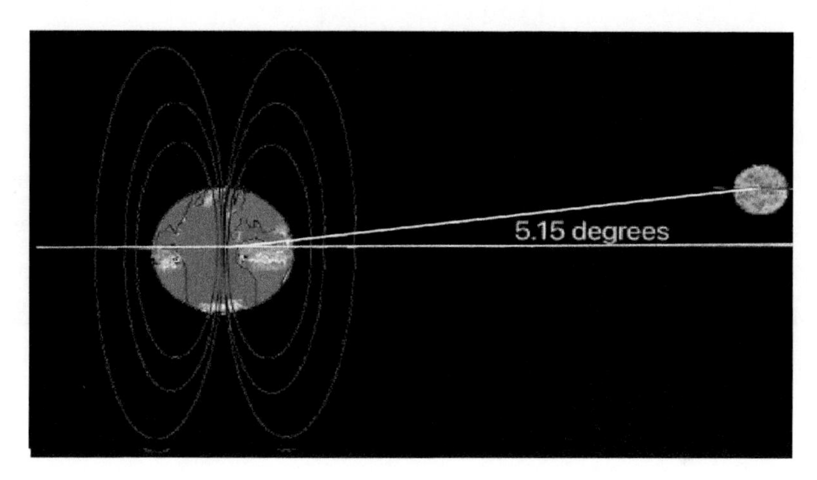

Figure 11.10 Simplified illustration of earth's magnetic field relative to the orientation of the moon and the gravitational plane of the earth's orbit.

The moon's position in its orbit around the earth is so precisely predictable that certain events such as lunar and solar eclipses can be calculated to the second as to when will occur and as to where they will visibly be seen from earth. The same applies to the observations of other planets and their own moons within the solar system.

My own specific theory is that the common force affecting orbits is caused by an attractive combination between high-frequency strings of energy in resonating harmony established in the instant of existence. This is the universal force, called gravity, that attracts all bodies in the universe. Its effect is far-reaching, extending to the edge of the known universe. More locally it can be observed in planetary motion, or as previously mentioned, the 'music of the spheres.' The sun's gravitational effect extends to beyond the known planets but causes the familiar ones to be held in orbit, circling the sun at different rates in accordance with the laws of physics.

A ball attached to a piece of string can be swung around at a rate mainly depending on the length of the string and the force that causes it to rotate. If we imagine that the sun's force of gravity is an attached 'string' of varying length, dependent upon distance from the sun, that provides the gravitational acceleration force, then the rotational rate can be calculated. The average distance of earth from the sun, called an 'astronomical unit', is 149 597 870.7 km (92 955 807.3 miles). By using this measure, a general rule is that cubing the distance in astronomical units and finding the square root of the answer gives the rate of orbit regardless of planetary size. For example, the orbital rate of Mars, which has a mean distance of 1.524 astronomical units, can be found by using the following formula:

$$\text{Orbit(years)} = \sqrt{(1.524)^3} = 1.87 \text{ years.}$$

Table 6 gives orbital data using this method and applying it to all the planets in the solar system. Pluto was downgraded from the status of a main planet to that of a minor one, but its orbit is also elliptical and exchanges places with Neptune as orbital distances cross over.

Planet	**Mean distance from the sun in kilometres × 1 000 000**	**Astronomical units based on distance of each planet's/earth's distance from the sun (AU)**	**Orbiting rate in earth years $=\sqrt{(AU)^3}$**	**Frequency value (Hz) YR × (1 / 31 558 118.4)**
Mercury	57	57 / 150 = 0.38	0.234	1.735×10^{-9}
Venus	108	108 / 150 = 0.72	0.61	1.9393×10^{-8}
Earth	150	150 / 150 = 1	1.00	3.1687×10^{-8}
Mars	228	228 / 150 = 1.52	1.87	5.9255×10^{-8}
Jupiter	779	779 / 150 = 5.19	11.697	3.7649×10^{-7}
Saturn	1430	889 / 150 = 5.93	14.44	4.5756×10^{-7}
Uranus	2880	2880 / 150 = 19.2	84.13	2.6658×10^{-6}
Neptune	4500	4500 / 150 = 30	164.31	5.2065×10^{-6}

Table 6 Planetary orbital distances from the sun and orbital rates around it.

The calculations in Table 6 work for all the planets regardless of size or mass. With these calculations, we can list the distances versus orbiting rates for all the major planets in the solar system and derive the frequency of orbital rotation for each one, as shown in the table. However, the force of gravity from the sun acting on the planets is not the only force. The table would be fixed and reasonably accurate but for the fact that other planets exert some attraction that slightly affects the orbit of each planet as it orbits the sun. This means that each orbit changes as other planets orbit the sun at the various distances and that the attractive force of gravity causes all others to vary slightly.

SOLAR ROTATION

Of course, many other frequencies occur that we are able to observe. The sun itself rotates, as can be seen when sunspots are visible on its surface, and even the sunspots rise and fall in intensity over a number of years. The sun has two accurately measurable rotation rates. One is called the solar rotation rate, and the other is called the synodic rotation. Solar rotation takes 24.47 earth days, but if a feature such as a sunspot is observed from earth, then the rotation takes 26.24 earth days to repeat itself. In fact, the sun's surface rotates at different rates depending upon latitude, with the highest number of days being around 36, at the poles.

Sunspots themselves occur cyclically with a recurrence of around 11 years and are also part of greater cycle that affects the sun's radiation. Evidence of this are the regular periods of heating and cooling of the earth's climate. At the time of writing, solar scientists are suggesting that a reduction in the heating effect called the Maunder Minimum may be about to be entered into. The last one was believed to have caused the mini-ice age when the river Thames froze over during the period from 1645 to 1715. These minimums would also appear to have a regular time period between peaks of activity.

REPETITIONS BEYOND THE SOLAR SYSTEM

Beyond the solar system, there are many repetitions to be found. The Milky Way is a spiral galaxy which we see on edge from the earth, and this is one of many billions of galaxies. In Chapter 2, I described galaxies as gravitational eddies. The term *eddy* means a swirling matter arising from influences or obstructions changing a smooth flow into a circular one that then forms a spiral. This can equally apply to free electrons in metal being caused to form eddies in any ferrous metal by induction from pulsing magnetic fields. Typical uses of this form of induction are found in steelwork smelters and also domestically in specially designed pans used with induction hobs. Pulsed magnetic fields from coils built into the hob induce these eddy current flows in the base of the pans. The swirling currents agitate the crystalline structure of the metal in a way that becomes thermal, heating up the metal. In the case of a smelter, using extremely strong pulsing magnetic fields, this is sufficient to provide enough heat to melt the metal and to keep it in a liquid form. This discussion may seem a galaxy away from what can be observed of our Milky Way, but as galaxies appear to be swirling masses of stars, there then is the connection. See Figure 11.11.

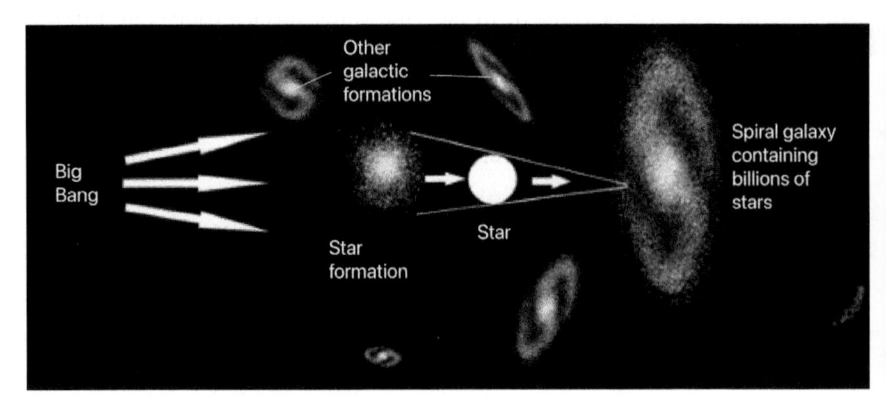

Figure 11.11 Big Bang → stars → galaxy formation sequence.

On the grander scale, the stars formed from the Big Bang also formed from swirling matter, and they too would have gravitational attractions to other stars and matter. This may have caused changes in direction in what may, up to a certain point, have been a straight trajectory from the Big Bang singularity. These deviations, repeated trillions of times, may have given rise to clusters swirling around a central 'hub', where the density would increase to such an extent that it forms a 'gravitational well', otherwise known as a 'black hole'. This gravitational attraction, plus the outward momentum, would give rise to clusters of many billions of stars looking like giant whirlpools in space. See Figure 11.12.

Figure 11.12 Andromeda galaxy.
©2022 Robin Barrett Astro-photographer

These clusters are attracted to the black hole whilst maintaining themselves as part of an expanding universe. To me the actual physical existence of spiral galaxies further proves that the energy of the universe forming the galaxies is expanding globe-like from a central point, the singularity.

The energy in the outward expansion is still there, not only in the expanding universe, but also in the rotation of the galaxies. Figure 11.12 is of the neighbouring galaxy to ours called the Andromeda galaxy, which is about 2.2 million light years distant. Our galaxy is 600 000 light years wide, so the width is equivalent to just under a third of the distance to Andromeda. Both Andromeda and the Milky Way galaxies are rotating about their respective central gravity wells and held in place by that attraction. It is also thought that 'dark matter' plays a part in the overall attraction process towards the gravitational galactic centre. For our galaxy, the rotational rate is given as 220 million years. This would suggest that there have been relatively few rotations since the Big Bang and that the eddies forming galaxies began very early on, from the starting point of existence. If the estimates of around 13.8 billion years are relatively accurate, then the maximum number of swirls possible to complete each full rotation would be $13.8 \times 10^9 / 220 \times 10^6 = 62.72$ rotations.

Although this discussion is based on speculation about the Big Bang, it shows that gravity became active from the very start and that gravitational attraction was a characteristic of all objects, from large formations to the smallest subatomic particle. This means that each object attracts every other object in the universe. This lends itself to a theoretical comparison of similarities between magnetic attraction and gravity since they may share a common link in the dynamics of the smallest subatomic entities making up all matter.

Magnetic fields arise from the movement of electrons, or charged particles, in a uniform direction. This means electrons caused to flow along a conductor or a specific molecular structure will form a field at 90° to the direction of flow as a continuous loop around and along that flow. Since the flow is from atom to atom, the field could be a manifestation of the energy released as each electron joins another atom which is missing an electron, caused by the flow, then neutralising the atom. The cumulative effect of the many billions of ordered re-joining atoms is the manifestation of a constant magnetic field at right angles to the flow. If the electromotive charge is increased, then the field formed increases proportionally as these 'leapfrogging' electrons become greater in number.

The linking between electrons displaced in atoms and then returning to their original orbits as an electromagnetic entity is similar to the formation of magnetic fields in general. Since there is a bipolar direction to all magnetic fields, whether static or alternating, that gives rise to attraction or repulsion, any fields nearby which are aligned in the same direction are attracted together. Conversely, any in the opposite alignments are repulsed, but not necessarily as one may think, since it is an attempt by one magnetic field to bring the other into a common alignment. This is by forcing it to physically flip to provide a matching attractive polarity. This effect is demonstrated by two magnets held close together with oppositely aligned fields. If one of the magnets is free to move and the other fixed, or larger, the second one will initially be repelled but will then flip and the two quickly snap together with the now aligned fields.

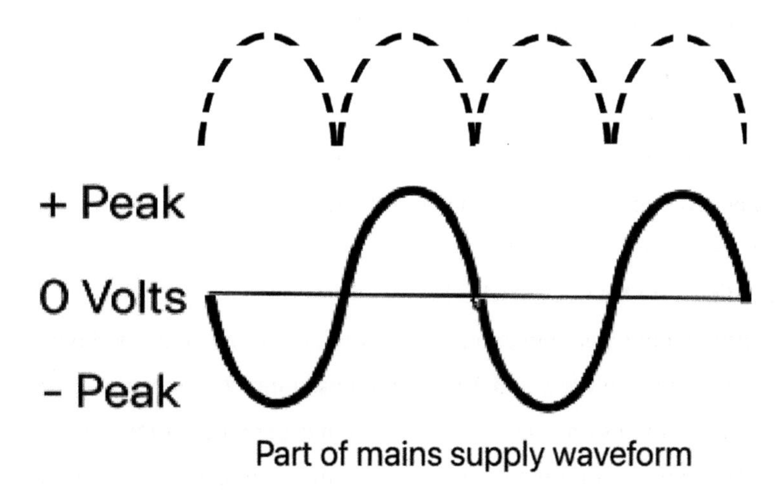

Figure 11.13 Gravitation electromagnetic analogy, 100 Hz buzz from a 50 Hz AC-supply sine wave. All frequencies in harmony try to pull together.

This attempt to physically align magnetic fields by creating a physical force is the basis for electric motors whose design takes advantage of this phenomenon. If all magnetic dipoles were aligned throughout the universe, then all ferromagnetic material within it would experience a force of attraction. This can further be demonstrated as follows: if the fields are generated electromagnetically by a solenoid and placed next to another energised solenoid, there will be a force of attraction between the two, provided that the fields are sympathetically bipolarly aligned. It would be the same even if the fields were oscillating all the time and fed in phase with the same frequency, although the strength of the attractions would vary with each cycle and be manifested as a mechanical buzz. See Figure 11.13.

APPLICATION TO POWER TRANSFORMERS

This is demonstrated by power transformers fed with 50 Hz from power grids. Each coil of the transformer is fed with the same AC supply, and each half of that supply waveform causes the magnetic field generated by each coil that is winding to be attracted to the adjacent coils. The overall effect is to produce a sound at twice the frequency of the supply. A 50 Hz sine wave has one positive and one negative peak for each cycle, as shown in Figure 11.13 but both peaks cause an attractive magnetic force. Since, overall, a 50Hz sine wave has 100 peaks per second, the buzz heard is at 100 Hz.

GRAVITATIONAL ATTRACTION

With gravity there does not appear to be a polarity or any repulsion but simply an attraction, but there could be gravitational bipoles that are aligned in all matter. This could be because gravity is a force generated from the very high-frequency vibrations running into many trillion, trillions of hertz of those minute particle strings that form all matter. This was alluded to in Chapter 2. Since gravity only attracts, it could be that all matter in the universe has particular gravitational stringlets within and that these are always vibrating in phase. They are the most numerous of all vibrational matter and also are the first and highest frequency formed in the instant of the Big Bang. Because it has this frequency component, this force of attraction forms gravitational fields that will attract all other fields formed from all other matter because they are all in the same phase and of the same frequency. This idea may suggest that gravity in itself can be broken down in the same way as photons by having the characteristics of a particle and a waveform, but on a scale far smaller and with a frequency far higher.

One may ask why photons are not attracted to each other in a similar way as gravity particles. My answer would be that photons are emitted as individual packets of energy from random electron activity within atoms and are always in transit and not all of the same frequency, whereas gravity particles are confined within the structure of matter, whose energy is at incredibly high frequencies and are all in the same phase and attracted to all others around, maintaining both subatomic and molecular cohesion. This is just my own theory, but it may be similar to others. It would show that there is a similarity between magnetism and gravity but there could be no direct connection because of the particular way these forces of nature are formed. Gravity could therefore be described in this way as a 'universal resonance' causing all things to be in 'particle harmony', attracted together, with the harmony being a peculiarity of a very specific frequency, as discussed in Chapter 2 and above. Although this gravitational attraction between all particles and mass would be minute to the extreme, when scaled up to the size of planets, the collective gravitational resonance becomes greater, right up to the point of the formation of stars and gravity wells such as black holes. The theories discussed above would suggest that if the frequency of gravity could be artificially generated, then, by phase adjustments, antigravity devices may be possible, bringing such science-fictional things into reality.

RATE OF FREQUENCIES FOLLOWING THE BIG BANG.

Ripples in the Ether II started with the Big Bang and a simple calculation taken from assumptions that the whole system oscillates from Big Bang to Big Crunch, creating another Big Bang and so on. I also suggested that all frequencies arose from this initial starting point for this cycle of existence. Looking at galaxies as a manifestation of those frequencies, with rotational rates taking millions of years as discussed above, we see that these rates provide the lowest frequencies currently detectable. With the Milky Way rotating once every 220 million years means one earth year is:

$$1 / 220 \times 10^{6} = 4.545 \times 10^{-9}$$

Dividing this by 31 558 118.4 seconds per year, we get 1.4403×10^{-16} Hz. This frequency value may be typical for those many billions of galaxies expanding throughout the universe that were also caused by the initial expansion and then were caused to rotate at the same rate as the Milky Way.

If proven, this then comes full circle to the range of frequencies within the time period that started with the Big Bang and goes up to the present, it being amongst the lowest repetitions in the known universe.

SPIRALLING GALAXIES AND THE FUTURE OF THE UNIVERSE

There may be another aspect that could be applied to these spiralling galaxies that may foretell the future of the whole universe. If, as is currently believed, each galaxy rotates around and is drawn to a black hole at its centre, then in some point in the future, all the material of the galaxy may be swallowed up by the black hole. If this applies to all the galaxies in the entire universe, then there would be countless billions of black holes left floating around. This would inevitably lead to gravitational attraction and the growing in size of those black holes as they are forced to join together. The conclusion to this scenario is that the universe would end up with just one massive black hole whose gravity would be so strong that it may collapse in on itself and return to that singularity discussed in Chapter 2.

Visualised typical black holes are usually represented as gravity gradients with objects being drawn into a vortex like a funnel, similar to water draining down a plughole. Artistic representation gives a similar impression. However, it is more usual to present these illustrations as just a one-sided point of view.

In most illustrations of black holes, material appears to be drawn into a bottomless gravitational pit. However, it is more likely that all matter is broken down into its basic fundamental parts, down to subquantum level, and gravitationally compressed back into a singularity. Figure 11.14 is a highly simplified illustration of gravitational attraction around a black hole.

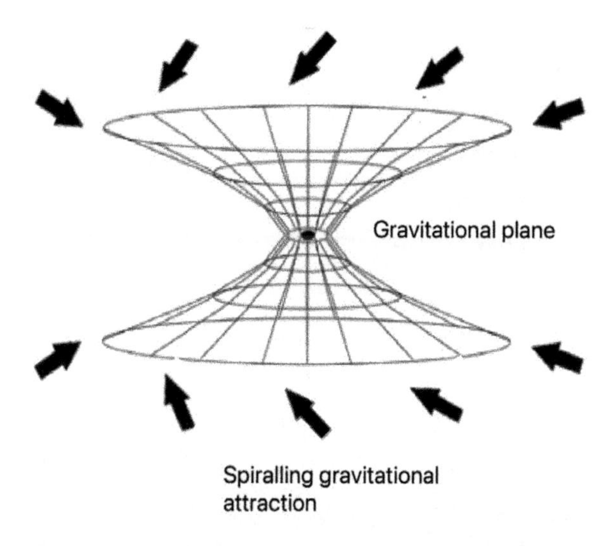

Figure 11.14 Simplified diagram of a gravitational well showing gravitational attraction from all sides towards the black hole.

The visible fact that a galaxy viewed edge on appears disc-like would suggest that there is a gravitational plane of attraction. This is similar to the situation with the sun in that all planets orbiting the sun are generally within the same plane. The same goes for the asteroid belt and the Oort cloud at the outer edge of the sun's gravitational influence.

The spiral disc-like formations form galaxies that bulge towards their centres. The reason for the greater number of stars circling in this region is probably that they are more densely populous as they approach the black hole and therefore start to bunch as they accelerate towards and around it, forming a spiral. The black hole is therefore attracting material from both sides of the galactic plane. The two spiral galaxies shown in figures 11.15 and 11.16 are perfect examples of the spiralling attraction towards the central gravity well or black hole.

Figure 11.15 Example of a symmetrical galaxy. This is the Whirlpool Galaxy
©2022 Robin Barrett Astro-photographer

Figure 11.16 Pinwheel galaxy.
©2022 Robin Barrett Astro-photographer

Ripples in the Ether II

Studying galaxies offers insights into the force that gravity exerts, extending many billions of miles/kilometres outwards from the black hole. As the photos show, galaxies are both beautiful and terrifying since the end of any galaxy is inevitable, falling into and becoming part of the black hole itself. It is just a thought of mine that if the Milky Way extends in its size about one third of the distance to our nearest neighbouring galaxy, Andromeda, then it is possible that the stars of the Milky Way nearest to Andromeda may be more under the influence of Andromeda's gravity than the stars at the opposite side. Likewise with the Milky Way's gravity affecting the nearest ones in Andromeda. The two galaxies are closing at an estimated fifty miles per second, so with this in mind, we are all probably under the gravitational influence of Andromeda. And the fact that the two galaxies appear to be on a collision course is evidence of a mutual attraction. So, with attraction theoretically emanating from the smallest oscillating strings at the possibly highest frequency with the minutest wavelength, there comes an attractive force that is the greatest force ever. This is felt throughout the entire universe and is responsible for forming the structure of the universe and its ultimate fate.

Chapter 12

CONCLUSIONS

Ripples in the Ether started out with discussions about the limitless dimensions of the void in which the universe exists and suggested that the concepts of infinity and eternity are beyond the natural understanding of the human mind. Although this final chapter summing up the frequencies that have been discussed is called 'Conclusions', such discussions on infinities means that nothing can ever really be concluded. This is because all our discussions are on the frequencies that manifest themselves as different forms of energy that really have no beginning and no ending and therefore go on forever. However, within the parameters that we are familiar with from the Big Bang to the present, there appears to be a continuation of that initial burst of energy where we are part of it as an ongoing kinetic manifestation. It is to me wondrous and drives a passion within me to contemplate existence and try to give meaning to it.

The late Richard Adams expressed this sentiment in his book *The Hitchhiker's Guide to the Galaxy* by posing a question to give the answer to 'life, the universe, and everything'. The great computer *Deep Thought* calculated the answer to be the simple number 42, this after millions of years of calculations. The book is fantastic and weird and is a very entertaining piece of fiction, but whatever the answer to the ultimate question is, it cannot be calculated because infinity has no number to base any calculation on either as a start or an end. As the book comes to its own conclusion, I hope that it has turned some of those frequencies and kinetic energy in the brains of the reader into further food for thought and contemplation.

This book has been on an excursion through time and the basis of existence manifested as frequencies, this from a beginning that we can just about understand, but not the beginning of everything. This is because eternity is an omnidirectional street in that it stretches back to and from infinity with no starting point and continues likewise forever to a time without end, both 'time eternal and distance infinite'. Our excursion through the frequencies that has occurred from a theoretically recognisable starting of time in this existence is limited to around the last 13.8 billion years, with theories that this may be part of a cycle that repeats itself at the lowest extreme of any frequency. From this arose all energetic forms of frequency right up to the spin of the smallest identifiable particles of an atom, the electron or perhaps the ejected neutrino. Perhaps beyond this are the frequencies of strings or stringlets that form the basis of all physical existence. This shows the range of frequencies that may be found in existence and the vast spectrum from the highest to the theoretical lowest available. In this book I have tried to group these frequencies from the very high to the lowest in order to cover the overall range. However, trying to arrange it in a neatly tied-up manner of cause and effect is not easy because of mixtures and overlaps of frequencies having different effects.

It would be an impossibility at this stage to give the frequency of the theoretical stringlets, as I called them, which I suggested gave rise to the many qualities and structures found within atoms. Perhaps, however, the highest frequency in the universe is the attractive synchronous frequency in all matter that gives rise to gravitational attraction. My theory of gravity may have been put forward already by others in perhaps a different form and perhaps presented using more scientific language. However, I have yet to come across any, apart from reference to the hypothetical possibility of quantum gravitons forming an overall basis for gravity, but none of these theorise on the fact that there does not appear to be any antigravitational force. Also, there are some still seeking to find a direct link between gravity and magnetism, but to my mind these are very different. I believe my explanation of harmonised gravitational vibrating strings, although very theoretical, may go some way to address the reason for this lack of antigravity.

COMPARING FREQUENCIES

As I have purposely not tried to put in order all frequencies discussed as such, there are exceptions. Frequencies within the electromagnetic spectrum can be, and were, discussed in order. Here, frequency comparisons and their effects can be made, and were discussed in descending order of frequency as the wavelength becomes longer. All these electromagnetic frequencies are subjected to one rule, namely the speed of light. The start of time and space is, in my opinion, tied to this speed. However, this occurred, it may or may not have a recurrence giving rise to a frequency of repetition that can only be estimated. All discussion on this is theoretical, but there are pointers that logically suggest an explosion having caused the universe to come into existence from a single point, and I believe that spiralling galaxies give some value to that suggestion and, to my mind, proof. Further pointers lie in the very high frequencies that gave rise to all forces and matter formation. Such frequencies could only be produced in the immeasurably short time frames after the start of the Big Bang.

The energy released from the Big Bang provided all energy around today. Energy is difficult to define, but we understand it as something that provides power, which in turn provides high frequencies manifesting as motion, light, and heat, the latter two being part of a wide spectrum where energy is transduced into photons. The potential energy tied up in every atom is a small piece of that moment of creation where the further manifestation of it is vibrational and further tied up in frequencies that then formed into all other types of matter in existence. The number of different frequencies possible is almost limitless. Since we have, in Chapter 2, established theoretically that all matter can be broken down into frequencies, we see that the number of possible forms matter can take is limited only by the frequencies possible and by combinations of those frequencies. So, now arriving towards the end of this book, I suggest that the highest frequency possible may be tied up with gravitational theory, for which my own theory has been discussed. It fits in with the idea of a frequency component nicely along with gravity waves having wavelengths measuring from a few kilometres to the diameter of the known universe. The range of frequencies that form gravity waves can be easily calculated because we now know they ripple through space, affecting everything around them, at the speed of light.

Calculating the frequency range from wavelengths, we discover the enormity of the scale is apparently from around 1 kilometre, being the lowest, to $2 \times 1.277 \times 10^{26}$ m, being twice my calculated photonic radius, assuming the universe is spherical. If the lowest wavelength is the width of the universe, hen:

$$2 \times 1.277 \times 10^{26}\ \text{m} = 2.554 \times 10^{26}\ \text{m},$$

then since $\lambda = c / f$, then $f = c / \lambda$, therefore:

$$3 \times 10^{8} / 2.554 \times 10^{26} = 1.746 \times 10^{-18}\ \text{Hz}$$

The shortest gravitational wavelength of 1 km gives a frequency of $3 \times 10^{8} / 1000 = 300$ kHz.

FREQUENCY OF GRAVITATIONAL WAVES

Looking at the very low frequency of gravitational waves, we know it is difficult to imagine how such a low frequency could be caused at one side of the universe and also why the gravitational scientists detect such low ones given that, as they transit at the speed of light, the effects would not be felt for billions of years. Such causes must therefore remain a mystery. However, thinking about the foregoing discussion, everything around us must exhibit a gravitational attraction. This means that every single thing that moves changes the relative gravity around it, including humans, all animals, and perhaps small variations in the passage of time. This further means that minute fluctuations are caused in the overall gravity field that radiates as gravity waves from the movement point, also at light speed. These are so minute in their effect as to be unmeasurable by any current methods known. This further suggests that although gravity waves share the speed of light as do photons, they are not like photons in that they are more a field of varying intensity rather than a particle, but have similar characteristics to electromagnetic radiation. Gravitons could be related to my stringlets idea and perhaps a manifestation of the very high frequencies peculiar to gravitational attraction.

The whole of existence, and contemplating all the factors that makes it perfect for life right here on earth is, for me, a wondrous adventure of the mind. The mystery of infinity is something I have difficulty with, not conceiving of it, but mindfully visualising something that never ends or equally has no beginning. The concept of the dynamics of existence, and that nothing is stationary even though objects may appear static, is also difficult to contemplate. Breaking the concepts down to their fundamental structures and particles reveals a dynamic world at the micro-micro levels of pulsating energy forming the frequencies, some of which have been included in this book. Life is a product of this dynamism, and our ability to contemplate it is a major part of it.

In Chapter 10, some of the discussion focussed mainly on the sound frequency band audible to the human ear, along with sensitivity and range. The simplistic discussion on the sound spectrum detectable by the human ear gave the frequency range from around 20Hz to 20 000Hz. In those frequencies are mixtures of harmonics and other different frequencies that give identity and meaning to the origins in an almost limitless number of combinations of a bass frequency, plus harmonics and others. To give meaning to this, consider a herd of sheep where newly born lambs are singularly identified by their mothers from their own unique bleats. To the human ear, they all sound the same, but within those sounds, along with many others occurring almost simultaneously from other lambs in the flock, the mother identifies and finds her own lamb. This is, perhaps, the same as or similar to human voice recognition in that a familiar voice is instantly recognisable. From within its patterns of frequencies, regardless of the actual words heard, are unique identifiers that are recognised by the human brain and are then allocated to the individual person from the 'library' of voices held within the brain's memory. The foregoing are examples showing that just classifying pure frequencies from 20 Hz to 20 000 Hz provides only a base of the almost limitless number of combinations that can be made from within those frequencies to give specific meanings.

Sound can provide a good analogy for other frequencies discussed in this book. With light, the most obvious are the colours seen in a spectrum or rainbow. Relatively few colours are emitted as light from within atoms, but the main three, red, blue, and green, can be mixed to provide an almost infinite number of combinations, something demonstrated by televisions, which can only ever display three-pixel colours. The same applies to our colour vision, but we see the glory of everything around us in full colour. If we take the example of photons from two different sources arriving at the same location, then we find they visually combine their frequencies but retain their individual identities. Take for example red photons at 630 nm illuminating a target viewed simultaneously with green photons at 530 nm and concentrated on the same target, which would be viewed as yellow at around 580 nm. This is the difference in frequency divided by two, subtracted from the highest colour frequency or added to the lowest colour frequency.

There are other factors depending upon intensities and the human eye's natural sensitivities to the intensity and hue of the yellow light viewed, but as can be seen from the spectrums, illustrated in Chapter 5, yellow sits very nicely between red and green. If yellow light produced in this way is then passed through a prism, the colours, and hence the photons, are split back to their original colours of red and green. In other words, photons do not change but what we see are simply combinations of them.

With transmitted electromagnetic energy, as with radio transmissions, the combination of two slightly different frequencies is demonstrated by two closely adjacent frequencies being received at the same time by a radio receiver.

Ripples in the Ether II

As a private pilot, I often hear a high-pitched whistle as two transmissions are picked up by my aircraft's radio at the same time, making neither of them translatable. The whistle is the very slight frequency difference between the two simultaneous incoming transmissions. With photons, the different colours arising are a construct of our brain from information received from our eyes from just three sensitivities of the rods combined with the cones.

Perhaps the biggest mystery for me is the origin of energy itself. Looking backwards through infinity, I wonder, has it always been there as oscillations of an unknown plasmatic form that occasionally pops into existence as Big Bangs? The enigma of discussing infinity, both in terms of the future and also of the past, if start or end points were to be suggested, then we, as humans, would pose the question of what occurred before those starting points. Conversely, what will happen after the end points? It is like a child asking why to every answer given by a teacher or parent after an enquiring question. If the answer does not satisfy, the child progresses further by asking why, why, why …? This can be annoying as each answer to satisfy this curiosity then fails. In adulthood, posing the question of why to existence and origins has to be more subtle, as any suggestion in scientific terms of an absolute answer can only be put forward as a theory when looking at infinities.

In writing this book, I have perhaps just scratched the surface of a seemingly very wide subject, but one where others are making new discoveries and theories are being put forward and added to all the time. An enquiring mind knows no limits, and terms such as *infinity* and *eternity* in this respect do not necessarily apply just to distance and time, but also to the number of variations of possible thoughts and contemplations about existence that may continue forever, or for as long as human minds evolve.

The 'music of the spheres' is to me more than just an expression relating to the predictable movement of planets but relating more to the music or frequency of life tied up in the harmonies of existence. The music continues and the harmonies hum everywhere from the time before time to the day after the end of time and may intrigue and stimulate the minds of thinking beings for as long as time exists.

At the beginning of this book, in Chapter 1, the start of an early paragraph was described as nonsensical because of the complexity of the subject. A reader casually glancing through and looking at this previous paragraph out of context may come to the same conclusion, but for those who have reached the end of this book by reading it throughout, then perhaps the latter part the penultimate paragraph just above now makes sense.

Appendices

NOTES ON SCIENTIFIC NOTATION

Both imperial and metric measurements have been applied in calculation examples used in *Ripples in the Ether*, particularly in reference to the speed of light.

This book is not meant to be mathematical, but out of necessity some calculations using extremely large or small numbers are simplified. For these we use scientific notation. However, for readers unfamiliar with scientific notation, which is used in various chapters herein, note that the superscripted number to the right of normal-sized numbers indicates the $10 \times$ numerical order, for example: $1 \text{ x } 10^3$ means $10 \times 10 \times 10 = 1000$, and $1 \text{ x } 10^6 = 1 \text{ x } 10 \text{ x } 10 \text{ x } 10 \text{ x } 10 \text{ x } 10 \text{ x } 10 = 1\,000\,000$. Similarly, $1 \text{ x } 10^{-3}$ superscripted means $(1/10) \times (1/10) \times (1/10) = 1/1000$, etc.

This simplifies extremely large or small fractions or cardinal numbers and makes multiplication or division of these numbers a simple exercise of either multiplying the main numbers together or dividing them by just adding or subtracting the small superscripted numbers as the case may be, for example:

$$(6 \times 10^{8}) \times (5 \times 10^{4}) = 30 \times 10^{12}$$

and for division:

$$(8 \times 10^{7}) / (5 \times 10^{3}) = 1.6 \times 10^{4}$$

For those with a negative sign in front of the superscripted number, the same applies, for example:

$$(3 \times 10^{-6)} \times (4 \times 10^{-7}) = 12 \times 10^{-13}$$

and for division:

$$(4 \times 10^{-6}) / (8 \times 10^{-3}) = 0.5 \times 10^{-3}$$

List of Figures

Note: Except where indicated, all diagrams in *Ripples in the Ether* were drawn by and are copyrighted by the author.

Images of Galaxies are provided by Astro-photographer Robin Barrett and are ©2022 Robin Barrett Astro-photographer.
We are grateful for his descriptions relating to these images, as shown below.

CRAB NEBULA FIG. 2.5

This nebula is the remnant of a violently exploding star, a supernova, about 6,500 light-years from Earth, and roughly 10 light-years in diameter. It was observed exploding by Chinese and other astronomers on July 4, 1054. The supernova was visible in daylight for 23 days and at night for almost 2 years!

In the late 1960s the Crab pulsar, thought to be the collapsed remnant of the supernova, was discovered near the centre of the nebula. The pulsar, which flashes in radio, visible, X-ray, and gamma-ray wavelengths at 30 times per second, provides the energy that allows the nebula to glow.

The orange filaments are the tattered remains of the star and consist mostly of hydrogen. The blue in the filaments in the outer part of the nebula represents neutral oxygen. Green is singly ionized sulphur, and red indicates doubly ionized oxygen. These elements were expelled during the supernova explosion.

60 images of 150 seconds each were taken in February 2022 over one night for a total exposure time of 2.5 hours.

ANDROMEDA GALAXY FIG 11.12

The Andromeda Galaxy is 15 billion-billion or 15,000 trillion miles from Earth. In light-years, which is the distance light travels at 186,000 miles per second, it's around 2.5 million light-years away. In other words, we're looking at the galaxy as it was 2.5 million years ago!

This is about the farthest thing in the Universe that we can see using just our naked eyes, visible only as a faint smudge.

The Galaxy is comprised over 1 trillion stars compared to 400 billion in our own Milky Way, with which it is on a collision course. But rest easy……they won't collide for another 4 billion years!

WHIRLPOOL GALAXY FIG 11.15

The Whirlpool Galaxy is a staggering 31 million light-years from Earth which is about 200 billion-billion or 200,000 trillion miles. This means that the photograph is as it appeared 31 million years ago! It has a beautiful face-on appearance, as seen from Earth, enabling us to see its distinct spiral structure and luminous core. When you consider that most of the stars we see are just tens or hundreds of light-years away, the distance to the Whirlpool Galaxy becomes phenomenal. It is Magnitude 8.4, making it about 400 x dimmer than the Orion Nebula and so much more of a challenge to photograph.

The spiral arms are packed with stars, gas and dust and make for the perfect conditions for new stars to be born, as that gas and dust coalesces, compresses and collapses to ignite star-birth. These hot young stars can be seen glowing bright blue along the arms, while older, yellower stars are seen glowing closer to the centre.

The Whirlpool Galaxy is easily recognisable due to its proximity to the yellowish dwarf galaxy NGC 5195 at the end of one of the Whirlpool Galaxy's spiral arms. This smaller galaxy is gravitationally interacting with the Whirlpool Galaxy, pulling the outermost stars in its spiral arm towards its own high-density core and creating further new star formation in the process

The images were taken in June 2022, and comprised over 90 exposures of 6 minutes with a total exposure time of 9 hours.

PINWHEEL GALAXY FIG 11.16

The Pinwheel Galaxy M101 is a classic spiral galaxy located 25 million lightyears away from Earth and is 170,000 lightyears across.

The galaxy is about twice the diameter of our own galaxy the Milky Way, and is thought to contain about 1 trillion stars.

The off-centre core and far-flung irregular spiral arms suggest that there may have been an interaction between this galaxy and other smaller galaxies. The image shows a predominance of scorching hot young stars glowing bright blue within the arms of the spiral galaxy. These regions are subject to intense bursts of star formation among the cosmic dust and gas, within molecular hydrogen clouds. These blue giant stars emit intense ultraviolet radiation that ionizes the hydrogen gas within their parent clouds, transforming them into bright reddish emission nebulae.

Ripples in the Ether II

The photographs were taken in February and March 2022, and comprised over 300 exposures of 5-6 minutes each giving a total exposure time in excess of 30 hours.

List of Tables

Glossary of Terms

alpha particles radiation particles consisting of a helium nucleus.

astronomical unit the mean distance between the earth and the sun.

barycentre the varying centre of gravity of the earth, the variation being due to the gravitational attraction of the earth–moon orbiting combination.

beta particles radiation composed of ejected electrons.

Big Bang theoretical instant of creation from a singularity. (*See* singularity.)

Big Crunch the theoretical coming together of all matter at a single point under the forces of gravity. (*See* singularity.)

chronon stated to be the shortest duration of time that can exist, and relating to the time required to traverse the classical diameter of an electron at the speed of light.

dark matter as yet unknown, undetectable material that is spread throughout the universe and estimated to make up 85 per cent of all matter.

decibel (dB) a logarithmic measurement scale of sound pressure variations, but used also in other comparative dynamic measurements having large variations.

frequency the rate in time at which an event repeats itself or cycles, returning to its starting state. (*See* hertz (Hz).)

gamma radiation the highest-frequency band of electromagnetic radiation emitted from within atomic structures.

hertz (Hz) the rate of repetitions or cycles of events occurring in a single second of time.

infrasonic mechanically generated vibrations or frequencies that are below the capability of human hearing to perceive.

photon energetic electromagnetic entity having the characteristics of being both a particle and a waveform.

singularity theoretical minute starting point of the Big Bang, from which everything originated.
speed of light the speed at which photons travel through a vacuum, approximated to be 300 000 000 metres per second (186 000 miles per second).
subsonic relating to objects travelling in the atmosphere or another medium that have velocities below that of the speed of sound through that same medium.
subunity frequencies frequencies that have repetition rates longer than one second. (*See* frequency *and* hertz.)
supersonic relating to objects travelling in the atmosphere or another medium that have velocities greater than the speed of sound through the same medium.
ultrasonic mechanically and naturally generated frequencies that are higher than those perceivable by humans.
X-ray electromagnetic radiation of a frequency band that overlaps the lower end of gamma radiation and is generated by highly accelerated electrons in a vacuum in collision with specific metals.

Hertz (Hz)	Cycles per second
kilohertz (KHz)	Hz × 1000 (10^3)
megahertz (MHz)	KHz × 1000 (10^6)
gigahertz (GHz)	MHz × 1000 (10^9)
terahertz (THz)	GHz × 1000 (10^{12})
petahertz (PHz)	THz × 1000 (10^{15})
exahertz (EHz)	PHz × 1000 (10^{18})
zetahertz (ZHz)	EHz × 1000 (10^{21})

Table 7 Frequency-related measurements for electromagnetic radiation (SI units).

Name	Symbol	Multiplier
centi	c	× 0.01 (10^{-2} m)
milli	m	× 0.001 (10^{-3} m)
micro	μ	× 0.000 001 (10^{-6} m)
nano	n	× 0.000 000001 (10^{-9} m|)
pico	p	× 0.000 000 000 001 (10^{-12} m)
femt	f	× 0.000 000 000 000 001 (10^{-15} m)

Table 8 Wavelength-related measurements (electromagnetic).

References

At the start of this book, I suggested that much of that found within would be original and from my own thoughts and not generally referenced to others. In general, I have kept to this, but in order to develop theories, one has first to acknowledge the work of others that provides the foundations on which to build. Standard order references are therefore readily included. The direct exceptions are those which have become internationally recognised in physics, such as Albert Einstein, Ernest Rutherford, Max Planck, Heinrich Hertz, and Michael Faraday, amongst many others. All these people have provided theories over the last two centuries that seamlessly intermesh, adding and taking from each other, to provide a fantastic body of knowledge and therefore need no dated reference as such. However, there are cases in modern research that cause contemplation and cannot go unacknowledged. The following list attempts to address this:

Alexandratou et al. (2002). Cellular responses to oxidative stress induced by light activation of zinc phthalocyanine. Conference paper. https://www.researchgate.net/publication/283504376_Cellular_Response_to_Oxidative_Stress_Induced_by_Light_Activation_of_Zinc_Phthalocyanine

Becker, R. O. (1998). *The Body Electric: Electromagnetism and the Foundation of Life.*

Braun et al. (2013). Negative absolute temperature for motional degrees of freedom. *Science*, 331, 6115: 52–55.

Britt, R. (2004). Universe 156 billion light-years wide. CNN. https://www.cnn.com/2004/TECH/space/05/24/universe.wide/index.html

California Institute of Technology (2016). www.ligo.catech.edu, accessed Jan. 2022.

Cancer Council of New South Wales (2020). www.cancer council.com.au

Chou et al. (2010). Optical clocks & relativity Science 329,1630;DOI:10.1126/science.1192720

Cremer, R. J., Penyman, P. W., and Richards, D. (1985). Influence of light on the hyperbilrubinanaemia of infants. *Lancet*, 7030: 1994–7.

Einstein, A., Podolsky, B., and Rosen, N. (1935). EPR paper: Can quantum-mechanical description of physical reality be considered complete?

Freire, O., Jnr. (1970). Quantum dissidents: Research on the foundation of quantum theory circa 1970. www.universe today.com

Garner, R. (2019). Mystery of universe expansion rate widens with new Hubble data. NASA.

Gotzsche, P. (2011). Niels fusion treatment for lupus vulgaris. *Journal of the Royal Society of Medicine*, 104, 1: 41–42.

Guth, A., and Steinhardt, P. (1984). The inflationary universe. *Scientific American.*

Hameroff, S. R., and Watt, R. C. (1982). Information processing in microtubules. *Journal of Theoretical Biology.*

Herrmann, N. (1997). What is the function of various brain waves? *Scientific American.*

Karu, T. I., and Afanas'eva, N. I. (1995). Cytochrome c oxidase as the primary photoacceptor upon laser exposure of cultured cells to visible and near IR-range light. *Proceedings of the USSR Academy of Sciences*, 342.

Krulwich, R. (2012). Which is greater, the number of sand grains on Earth or stars in the sky? -7 pm

L'Annunziata, M. (2007). *Radioactivity: Introduction & History,* Amsterdam: Elsevier.

Lardner, B., and Lakim, M. (2002). Animal communication: Tree frogs exhibit resonance effects. www.semanticscholar.org/paper/animal

Lesgourgues, J., and Verde, L. (2019). *Neutrinos in Cosmology.*

Matsumoto Y et al (2018) Visualising peripheral arterioles and venules through high-resolution and large-area photo acoustic imaging. Sci Rep 8, 14930.

McCaig, C., et al. (2005). Controlling cell behaviour electrically: Current views and future potential. *Physiological Review*, 85, 3: 943–78.

Petrakis, L. (2004). Ancient Greeks and modern science: Who discovered the heliocentric system? PhD thesis, New York: Herald.

Roemer, O. (1676). *From Cosmic Horizons.*
Sahu, S., and Andyopadhyay, A. (2013). Energy levels in microtubules. *Biosensors and Bioelectronics*, 47.
Saslaw. (1991). Black holes and structure in an oscillating universe. Nature https://www.nature.com/articles/350043a0

Schwarz, J. H. (2000). *Introduction to Superstring Theory https//archive.org/ details/arxiv-hep-ex0008017*

Siegel, E. (May 5th 2016). Why does gravity move at the speed of light? https:// www.Forbes.com

Somerville, D. (2018). A guide to phototherapy.

Sutter, P. (2020). What happened before the Big Bang? Live space science April 20, 2020

Soter and Tyson. (2000). New Press, American Museum of Natural History.

Universe Today. www.universetoday.com/36302/atoms in the universe

Townes, C. (2003). The first laser—Century of nature: Twenty-one discoveries that changed the world.

Turner, J., and Hode, L. (2002). *Laser Therapy: Clinical Practice and Scientific Backgrounds.* Prima.

Turok et al. (2004) M. Theory model of a Big Crunch/Big Bang transition. High Energy Physics-Theory Cornel University

Wood, C. (2019). A simplified explanation and brief history of string theory.

Young, (1965). The mystery of matter.

About the author

David C. Somerville has had a varied career leading to a wealth of experience. He served in the Royal Air Force, training as an air wireless mechanic, and later gained a commission in the volunteer reserve. He went on to become a senior computer-engineering analyst before earning a degree in physical sciences and education and a doctorate in orthopaedic research.

After several years as a university lecturer, he became a consultant and freelance lecturer, setting up his own business designing electrotherapy equipment. He is semiretired and a private pilot. David's lifelong involvement with the application of frequencies throughout his career in the Royal Air Force and industry, along with his realization that all things are full of resonances inspired him to write this book about space, time and the underpinning frequencies and resonances affecting everything around us. His three previous books were related to the underpinning science of therapy equipment.

Notes

Notes

Notes

Notes

Notes

Notes

Notes